SECRETS OF
EARTH AND SEA

BY

I0390808

Sᴉʀ RAY LANKESTER
K.C.B., F.R.S.

WITH NUMEROUS
ILLUSTRATIONS

SECRETS OF EARTH AND SEA

DIMETRODON GIGAS,
AN EXTINCT LIZARD, SEVEN FEET LONG

PREFACE

THE present volume is, like its predecessors, "Science from an Easy Chair" (Series I and Series II) and "Diversions of a Naturalist"—mainly a revision and reprint—with considerable additions—of articles published in daily or weekly journals. The first chapter appeared originally in "The Field." The Chapters VI, XX, XXI, and XXII were published in the "Illustrated London News," under the title "About a Number of Things." The rest are some of the articles which, as "Science from an Easy Chair," I contributed, during seven years, to the "Daily Telegraph." That, to me very happy, conjunction was, like so many other happy things, necessarily interrupted by the Great War.

One result of that terrible cataclysm is that not a few thoughtful writers have been led to deny the existence of what they call "Progress," meaning by that word the development of mankind from a less to a more complete attainment of moral and physical well-being. The question raised is obscured by the arbitrary use of the word "progress," since by it any movement from point to point—whether advantageous and desirable or the reverse—is described, as, for instance, in the familiar titles given by Bunyan to his book "The Pilgrim's Progress" and by Hogarth to his pictures "The Rake's Progress." Those who to-day despair of man's future limit their outlook on the past to the conventional history of some three or four thousand years. The only solid ground upon which we can base the supposition that mankind has moved from a less to a more complete attainment of moral and physical well-being and will continue to do so, exists in the ascertained facts of the past history of living things on this Earth, and of man since his earliest emergence from among the man-like apes made known to us by his stone-implements and fossilized bones. That there has been a development from lower, simpler structure to higher, more complex, more efficient structure is demonstrable, and so is the proposition that there has been in the human race a continuous development in the direction of increased adaptation to the conditions of social life and an increased control by man of those natural agencies which he can either favour when conducive to his prosperity, or on the other hand can arrest when inimical to it. "The continuous weakening of selfishness and the continuous strengthening of sympathy" (to adopt the words of the American philosopher, Fiske) are, in spite of numerous lapses and outbursts of savagery, patent features of the long history of mankind. We have no reason to doubt their continuation, whilst at the same time we must be prepared for and accept, without desponding, the ups and the downs, the disasters as well as the triumphs, which inevitably characterize the natural process of evolution. One thing, above all others, we as conscious, reasoning beings can do which must tend to the further development and security of human well-being: we can ascertain ever more and more of the truth, or in other words, "that which is." We can discover the actual conditions of natural law, under which we exist and promote the knowledge of that truth among our fellows. To do that which is right, we must know that which is true. To act rightly, we must know truly.

We possess, a vast heritage of knowledge handed on to us in tradition and in writings from our father-man in the past. But there are yet immense fields of knowledge to be explored and yet a greater task to be accomplished in spreading the knowledge which we possess, and in persuading all men that it is their right and their duty to acquire it and to enjoy the power and the pleasure which it gives. All must also help, directly or indirectly, in the making of new knowledge. Whilst mankind is still so backward in knowledge and the worship of wisdom, it is idle to indulge in despair of the future. A chief way to increased welfare is still open and untrodden.

These are big speculations and problems with which to preface a small book. But I am content to offer the small book as a contribution, however restricted, to the spread of a desire for further knowledge of the things about which it tells—a possible incitement to serious study of some one or other among them.

E. RAY LANKESTER

June 2nd, 1920

EXPLANATION OF THE
FRONTISPIECE

THIS plate shows the restoration of the extinct lizard, Dimetrodon gigas (Cope), lately made by Mr. Charles W. Gilmore of the United States National Museum, by whose kind permission it is here reproduced from the Proceedings of the U.S. National Museum, vol. 56, 1919. It is based upon the study of a very fine skeleton and some hundred bones of allied species, collected by Mr. Sternberg from "the Permian formation" exposed in the vicinity of Seymour, Texas, U.S.A. It is selected for illustration here because its most striking feature—the high dorsal fin-like crest along the middle of the back formed by the elongation of the neural spines of the vertebræ—is a puzzle to the conscientious Darwinian. Professor Case says of it: "The elongate spines were useless, so far as I can imagine, and I have been puzzling over them for several years. It is impossible to conceive of them as useful either for defence or concealment, or in any other way than as a great burden to the creatures (terrestrial non-aquatic animals) that bore them. They must have been a nuisance in getting through the vegetation, and a great drain upon the creature's vitality, both to develop them and keep them in repair." The reader is referred to pp. 127, 128, where a brief discussion of such exuberant growths will be found. The excessive growth of the median fins in the fish Pteraclis allied to the Dolphin which displays changing floods of surface colour as it dies—and in the Australian Blenny called Patæcus—both figured on p. 130—should be compared with that of the strange crest of the grotesque Dimetrodon.

SECRETS OF EARTH AND SEA

CHAPTER I

THE EARLIEST PICTURE IN THE WORLD

IN Figs. 1 and 2 on the next page a cylindrical piece of the antler of a red deer is represented of half the natural size. On it are carved by in-sunk lines certain representations of animals. It was found in the cavern of Lortet, near Lourdes, in the department of the Hautes Pyrénées, in the south of France, together with many other remains of prehistoric man. This cavern was excavated and all its contents of human origin carefully preserved by M. Edouard Piette in 1873 and the following years. Drawings of this and other remarkable carved pieces of bone and antler, many in the form of harpoon heads, and of small chipped flint implements, all found in this cave, were published by him. [1] He excavated also several other caverns with great care, and his collections were bequeathed by him on his death to the great Museum of National Archæology at St. Germain, near Paris, where I have had the advantage of studying them.

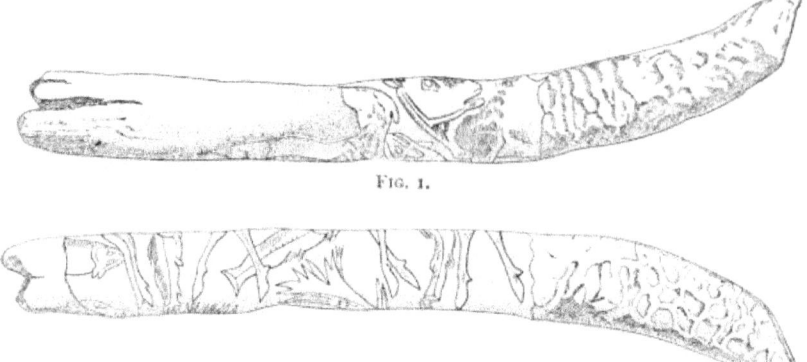

FIG. 1.

FIG. 2.

FIGS. 1 AND 2.–Engraved cylinder of red-deer's antler, from the
Azilian (Elapho-Tarandian) horizon of the cavern of Lortet.
Drawn of a little more than half the actual size of the specimen.

The age assigned to this carving is that called by Piette
"Elapho-Tarandian." At this period the reindeer
(Tarandus), which previously abounded, is giving place to
the red deer (Elaphus). The layer in which this carving was
found belongs to the latest of the Palæolithic cave deposits,
and was followed by a warmer period, in which the red
deer and the modern fauna entirely replaced the old fauna
of the Glacial period. The deposits in Pyrenean caves of
the Elapho-Tarandian age are characterized by an
abundance of large flat harpoons serrated on both sides. In
this latest horizon of the Reindeer period the art of
engraving in outline on bone and stone had attained the
highest pitch of excellence which it reached in the
prehistoric race of South-West Europe.

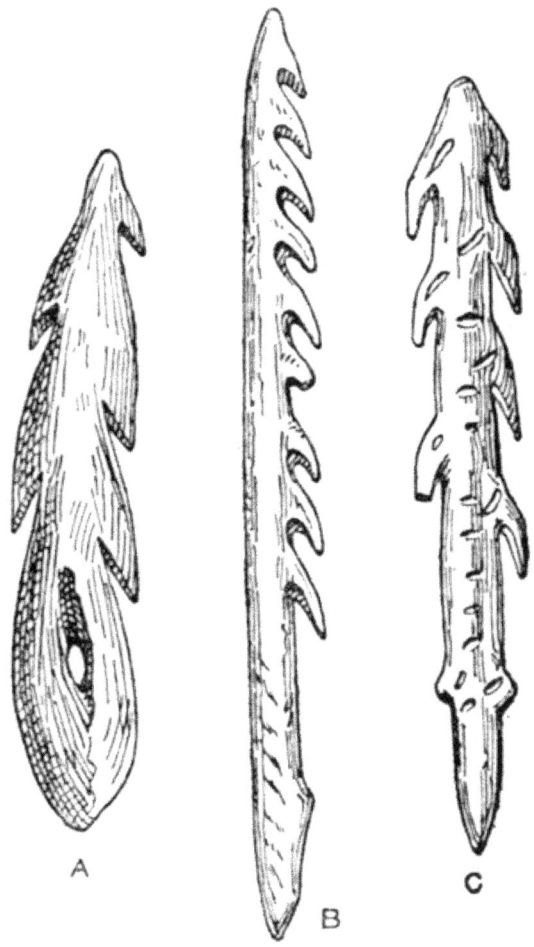

FIG. 3.–*A*. Perforated harpoon of the Azilian or Red-Deer period, made from antler of red deer, found in quantity in the upper layers of deposit in the cavern of the Mas d'Azil (Arriège). *B* and *C*. Imperforate harpoons or lance heads made from reindeer antler of the Magdalenian period (Reindeer epoch). *B* from Bruniquel Cave (Tarn-et-Garonne). *C* from a cavern in the Hautes Pyrénées. Same size as the objects.

A very natural tendency among those who hear from time to time something of what is being discovered about primitive man is to confuse all the periods and races of prehistoric man together, and so picture to themselves one ideal "primitive man." My friend Mr. Rudyard Kipling does this, although it would be no further from a true conception were he to blend his ancient Britons, his Phenicians, his Romans, his Saxons, his Normans, and a few Hindoos into one imaginary man and represent him as taking a coloured photograph of the Druids of Stonehenge on a piece of Egyptian papyrus. Here is Mr. Kipling's vision of primitive man:

Once on a glittering icefield, ages and ages ago,
Ung, a maker of pictures, fashioned an image of snow.
Later he pictured an aurochs, later he pictured a bear–
Pictured the sabre-tooth tiger dragging a man to his lair–
Pictured the mountainous mammoth, hairy, abhorrent, alone–
Out of the love that he bore them, scribing them clearly on
    bone,
Straight on the glittering icefield, by the caves of the lost
    Dordogne,
Ung, a maker of pictures, fell to his scribing on bone.

The fact is that several prehistoric races have succeeded one another in Western Europe during the immensely long period—amounting to hundreds of thousands of years— during which man existed before the dawn of history. The "lost" or "prehistoric Dordogne" was like the present historic Dordogne in regard to the fact that many races and dynasties successively held possession of it and left their work in its soil and caves.

Passing back through the historic age of iron and the sub-historic age of bronze, we come to a time, about four thousand years ago, when there were no men in the west of Europe who made use of metals at all, although, for a thousand or two years earlier, men were using bronze and copper in the East. European races immediately before the first use of metals made beautiful implements of stone (chiefly flint), and finished them by grinding and polishing them. These men are spoken of as Neolithic men, or men of the Neolithic period. They had herds and cultivated crops, and they built after a fashion rough houses in wood and tombs and temples with great slabs of stone. They made pottery and woven cloth. The animals and plants of Europe were the same in those late prehistoric times as they are to-day. The Lake dwellings of Switzerland belong to this epoch and yield us their remains as evidence. The men had very nearly the same set of domesticated animals as we have to-day, but they had no skill in carving outlines of animals. Their only decorative work consisted of parallel lines, straight or in zigzags or in circles, graven on the great stone slabs which they erected.

We can trace them back to some seven thousand years B.C. and then comes a huge gap—we do not know how many thousand years—in our evidence as to what was going on in this part of the world. We find convincing proof that before this interval the climate was much colder than it is to-day, and that the land surface of Europe was in many respects very different from what it became later. Britain was continuous with the Continent. There were in that remote period human tribes spread over the less frigid valleys of Europe. They had no fields, no herds; they fed on the roasted flesh of the animals they chased and on the fish they speared, and on wild fruits and roots. They dwelt chiefly, if not wholly, in caves, probably also in skin tents, but they did not build either in wood or in stone. The age which we thus reach is called the Palæolithic, or "ancient" Stone age, because men made use of stone, which they chipped into shape, but, unlike the Neolithic people, never polished it. We find enormous numbers of these rough or Palæolithic stone implements both in caves and in the gravels deposited in the ancient beds of rivers. They are so abundant as to prove the existence of a very considerable human population in the remote ages when they were fashioned and used. The changes which have taken place and the time involved since some of these Palæolithic implements were made and used may be guessed at (but cannot be definitely calculated) from the fact that the beds of the rivers which formed the gravel terraces in which they are found in England were, in many cases, from one to six hundred feet above the level of the present rivers. The land surface has risen and the rivers have simultaneously excavated deep and wide valleys leaving terraces of gravel high up on their sides. These show where the rivers once flowed. The vastness of the excavation of the valley from the level of the old river bed 600 ft. up on the sloping hill-side to its present low-lying bed in the floor of the valley—gives us some measure of the time which has elapsed in the process.

No one can tell, at present, the limit in the past of Palæolithic man. The period of time over which his existence extended, as indicated by the trimmed flints undoubtedly made by human workmanship, is a matter of hundreds of thousands of years. In Western Europe races came and went, succeeded one another and disappeared, either migrating or absorbed or more rarely destroyed by the later invaders. Naturally enough, in the later deposits of rivers and in the higher layers of earth and limestone cake which fill many caves to the depth of 30 or 40 ft. we find the remains of man's workmanship more abundantly than in the older deposits.

We can broadly distinguish in the Palæolithic epoch three (perhaps four) periods, separated by the occurrence of great extensions of the northern or arctic ice cap of such a volume as to cover North Europe and North America, and the simultaneous extension of the glaciers of the mountains of Europe. This period of the alternating extension and retreat of the great northern glaciers is known as the Glacial period, or Ice Age. The *latest* Palæolithic men are subsequent to it—that is, post-Glacial.

We can distinguish several successive ages of these post-Glacial Palæolithic men, altogether distinct from and anterior to the Neolithic men. In the earlier of these ages many of the great animals of the Glacial period—now extinct or withdrawn to other regions—still survived in Europe. The mammoth survived, but was fast dying out in the south and centre of France, and we find its outline scratched on ivory and on bone by the early post-Glacial men. The lion still survived in Europe, also the hyena, the bear and the rhinoceros. The reindeer seems to have been especially abundant, and to have been associated with the men of this period. The horse was very abundant, and was largely eaten by the earlier post-Glacial people. From the first these men show extraordinary artistic skill, and have left in their caves many carvings on ivory, bone and stone. In the oldest deposits of the post-Glacial age the carvings are complete all-round sculptures of small size or carvings in low relief, all of rough primitive workmanship. Larger life-size sculptures in rock are also found. In later deposits we find better sculpture and also engraving on flat pieces of bone and ivory, and also on stone. This art persisted, and attained its greatest perfection in the latest deposits of all in which the work of Palæolithic man is found. The reindeer persisted through this post-Glacial period (hence often called "the reindeer period") until the gradual increase of temperature and change of herbage and forest led to its migration northwards and to the relative abundance of the red deer. It is to this latest period—the Elapho-Tarandian of Piette—that the engraved antler figured here (Figs. 1 and 2) belongs.

At an earlier stage of the post-Glacial period men hunted the bison and other large game in the north of Spain and made coloured drawings of them on the roofs and walls of their caves, drawings which have been copied and preserved: whilst the mammoth, the rhinoceros, the cave lion and bear still inhabited south central France and are pictured on the walls of caves in that region—as described in Chapter II. Later we lose all trace of Palæolithic man and his wonderful artistic skill. He seems either to have migrated or to have been absorbed in the immigrant Neolithic race—a race singularly devoid of any tendency to artistic sculpture or engraving.

The skeletons and skulls of the men of the Reindeer period, or post-Glacial Palæolithic men, have been discovered here and there. They indicate a fine, tall people with well-shaped skulls and jaws, comparable to the nobler modern races. It is convenient to call them Cromagnards, since good skulls of the race have been described from Cromagnon, in France. There is evidence (from skulls) that another race (the negroid so called "Aurignacians") preceded and coexisted to some extent in Western Europe with them, but we have, at present, no evidence as to whence or how the Neolithic race or the Cromagnard race or any of their predecessors came upon the scene!

When we go farther back and reach the actual Glacial period we find a very different state of things. The men who then existed in the caverns are called the Neander men. They were a short, bandy-legged, long-armed, low-browed people, great workers of flints. They had the use of fire, and contended with hyenas and bears and lions for the occupation of their caverns. In their day—the day of European glaciation—the mammoth was in full occupation of the pine forests on the edge of the glaciers. But the Neander men made no sculptures, or carving, or engravings. The gap between them and the Cromagnon men is much greater than that between an Australian black fellow and an average Englishman; indeed, the difference is properly expressed by regarding the Neander man as a distinct species—Homo neanderthalensis.

Passing again farther back over an immense period of time, we find Europe warm again; the glaciers have (for a time) gone or retreated far up the mountains but are found in extension again at a still earlier date. An inter-Glacial set of animals is now found living in a comparatively warm climate in Western Europe. Another elephant (Elephas antiquus) is there (not the mammoth), and another rhinoceros (not the woolly rhinoceros of the later Glacial period); the hippopotamus flourished then in Europe and swam in the Thames and Severn, and there too, at last is the sabre-toothed tiger, which did not exist at all at a later period! Now was the time when a man, if he could, might have "scribed" the image of a sabre-toothed tiger on a piece of bone, but, so far as we know, he did not and could not. This was ages before other succeeding men walked "on glittering ice fields," and they, in turn, were ages earlier than the artistic Cromagnards of the Reindeer period.

The presence of men in the warm inter-Glacial times in Europe is proved by the association of rough but undisputed flint implements with the inter-Glacial animals and by the discovery of a most interesting human jaw (chinless, like that of the Neander men) in what is held to be a præ-Glacial deposit at Heidelberg. We have very little knowledge of Glacial and præ-Glacial man except well characterized flint implements and two skeletons, some detached limb bones, four or five jaws, and as many skulls. [2] But of post-Glacial Palæolithic man we know the skeletons of the Cromagnard race, their sepulture, their decorative necklaces, and their bone and ivory carvings and engravings, and the coloured rock paintings and other work of earlier races (the Aurignacians, and others) belonging to successive epochs or eras, which have been discovered in caves in France, Spain, Belgium, and

Austria. It was long after them that the Neolithic people appeared.

The preceding remarks will have made it clear that the engraved antler here figured was carved by a man who was not really at all primitive, although he lived probably between twenty and fifty thousand years ago. It will also have been made clear that hundreds of such engravings, more or less fragmentary, are known. Some are very skilful works of art, others of a much inferior quality. Many, however, show an astonishing familiarity with the animal drawn and a sureness of drawing which is not surpassed by the work of modern artists (see Chapter III). The interest of the particular engraved antler which I am describing is that it is the only carving of its age as yet discovered which is more than a drawing or sculpture of a single animal. It is a "picture" in the sense of being a composition. It is not, it is true, painted—it is engraved; but being a composition it is entitled to be called "the earliest picture in the world." Let me describe it a little more fully with the help of the illustrations.

The engraving has been made on a long cylindrical piece of the red deer's antler. It can hardly be considered as decorative, since the figures of the animals do not show as such on the cylindrical surface (Figs. 1 and 2). Pieces of antler, bone, and ivory carved with spiral scrolls and circles which are really decorative and effective as decoration are found in these caves (Fig. 29). But often such pieces as the present are met with. It has been discovered by French archæologists that the true intent of such engravings may be rendered evident by rolling the cylinder on a plastic substance (soft wax or similar material), when the drawing is "printed off" or "developed" as it is termed. A great number of such line engravings have been thus printed off or developed, and plaster casts made from the flat impressions are preserved in the museum of St. Germain, the engraved lines being rendered obvious by letting them fill with printing ink. They often give us in this way a "printed" drawing of remarkable accuracy and artistic quality. The rolled-off print of our specimen is shown in Fig. 4. The cylinder has been damaged by time, but the print shows, more or less completely, a vigorous outline drawing of three red deer, with six salmon-like fish placed in a decorative way above them and between their legs. Two lozenge-shaped outlines (above the larger stag) are held by good authorities to be the signature of the artist. The group of deer is represented in movement. The largest stag is on the right; his hindquarters are broken away by injury to the cylinder. He is commencing to advance, and turns his head backwards to see what is the thing which has alarmed him and his companions; at the same time his mouth is open, and he is "blowing." The second stag is a younger and smaller animal, and is retreating more rapidly. The cylinder is damaged so that, although all the four legs of this second stag are preserved, the head and neck are gone, though the points of the antlers are preserved. The same damage has removed all but the hind legs of the still younger animal who heads the group. The beauty of the drawing of these hind legs and the extraordinary impression of graceful, rapid movement given by their hanging pose, side by side, is not surpassed, even if it be equalled, by the work of any modern draughtsman. It is clear that the youngest and smallest member of the group is, as is natural, the most timid, and that he has sprung off with a sudden bound on the occurrence of the alarm from the rear, which is setting the whole group into motion with increasing velocity as we pass from right to left.

FIG. 4.—Rolled impression or "development" of the engraving on the Lortet antler.

FIG. 5.—Restoration (or completion) of the engraving on the Lortet antler, as now (1919) suggested by the writer (E. R. L.).

The "printed-off," or "unrolled," or "developed" picture given in Fig. 3 is an exact reproduction of a copy of the cast made and preserved in the Museum of National Antiquities at St. Germain, for which I am indebted to my friend M. Salomon Reinach, the distinguished archæologist who is the director of that museum. It is reproduced here, a little larger than half the size of the original, as are the representations of the carved cylinder itself (Figs. 1 and 2). In Fig. 4 we have my attempt to restore the damaged portions of the design and to present it as it was when the Palæolithic man completed it some 20,000 years ago.

I will return to the question of the correctness of this restoration, but before doing so I wish to mention some extremely interesting points as to the probable use of the cylinder of stag's antler and the purpose of the carving around its axis. In the first place, this and a few other of the pieces of carving of the post-Glacial period were certainly the work of highly gifted and practised artists. It is obvious that this work is far superior both in conception and execution to the more or less clever, often grotesque, carvings and paintings made by modern savages or simple pastoral folk. There is no reason to suppose that the Cromagnards, or men of the post-Glacial or Reindeer period of West Europe, differed from modern races in being universally gifted with artistic capacity. This engraving of three stags is almost certainly the work of a man who belonged to a family or guild of picture-makers who had cultivated such work for centuries and handed it on from master to apprentice. This design is probably one which had been perfected by many succeeding observers and draughtsmen. Its sureness of line and vivacity of movement are not the outcome of the sudden inspiration of an untutored savage, but are the result of the growth, cultivation, and development of artistic perception and the power of artistic execution in successive generations.

It seems in the highest degree improbable, if not impossible, that so excellent a drawing as this should have been cut on the cylindrical piece of antler by an engraver who never saw the flat or rolled-off impress of his design. One is driven to the conclusion that he must, as he worked on the bone, have taken an impress of the growing picture from time to time, using probably animal fat and charcoal as an "ink" and printing on to a piece of prepared skin or on to a birch-bark cloth. How otherwise could he have made his engraving so truly that when, ages afterwards, we print it off the cylinder, we are astonished and delighted by its perfection of design and execution? If this be once admitted—namely, that the artist tested and checked his work by printing it off as he proceeded with it—we gain what appears to me to be the probable solution of the question which has been largely debated, "For what were these carved cylinders or rods used?" Those which are simple cylindrical rods, such as the present one, must be distinguished from others which have one or more circular holes bored in them and others which are curiously bent at an angle. Such specimens are often carved with small unimportant ornament, not requiring development or printing. They as well as the present class have been spoken of as "wands of authority" and "sceptres"; some are considered to be arrow straighteners; others have been supposed to be "divining rods" or "rods of witchcraft"; whilst one of those discovered by M. Piette (others similar to it are known) has been regarded as a "lance thrower" or "propulsor" (such as modern primitive races use), having a notch at one end upon which the lance to be thrown is made to rest. The latest suggestion as to these notch-and-hook-bearing rods, is that they are large crochet hooks used in making nets. It has also been suggested that some of these carved rods were used as "fasteners" of the skins used as clothing.

I venture to suggest that the elaborately carved cylinder which we are considering and others bearing similar carvings, which only show up when a printing of them is taken, were used by the men who made them for this very same "printing" as an end in itself. The picture could be thus impressed on skins, birch bark, and other material. This race was thoroughly familiar with the use of paint formed by mixing grease with charcoal (to produce black), red ochre (to produce red), yellow ochre (to produce yellow), and some preparation of limestone or chalk (to produce white). Coloured pictures representing animals of the chase, coloured with red, yellow, white, and black and outlined by engraving, have been discovered on the rock walls of the caves used by them. Such pictures are found of relatively early as well as of late date within the post-Glacial Palæolithic period (see Chapter III). The rock picture of a single animal is usually from two to five feet long. People who could make those coloured designs and who could draw and compose so admirably as the author of the "Three Red Deer" would have desired to "roll off" and to possess printings of their favourite representations of animal life, whilst we must admit that their skill and ingenuity was assuredly equal to the task of so printing them. If this carving of the "Three Red Deer" were never printed it could not have been executed in the first place, nor seen and admired when completed. If even only half a dozen or a dozen impressions were taken from it for ornamenting the skins or other material used by a chief, or a wizard, or a woman, its production becomes intelligible. It is true that there is nothing known as to the use of such printing from a cylinder among existing primitive people, but it is known in very early times (4500 B.C.), since cylindrical seals were used by the Babylonians. Elaborately grooved blocks used for printing on cloth are known from Fiji and Samoa, and the mere practice of printing on to a flat surface is common enough among savage races in regard to the human hand, impressions or prints of which obtained by the use of a greasy pigment are found upon rocks or stones. Sometimes prints of the hand or fingers are taken in clay.

We must not, however, forget that the primary purpose of savage and primitive mankind in making images or engravings of animals is that of influencing the animals by witchcraft or magic, as has been urged by Reinach. From such magic-working drawings the art of savages has gradually developed just as religious figures and designs have been the initial motive of historic European art.

It seems in any case fairly certain that the artist who engraved our picture of the three deer on to the stag's antler must have worked from and copied a completed flat drawing, and probably printed it in some way on to the prepared antler before engraving its lines thereon and also checked the work, as he proceeded, by successive trial printings or "proofs" on to a flat surface. It is possible though it does not seem very probable, that the drawing was thus committed to perpetual invisibility on a cylindrical rod—for the purpose of exercising "magic" with that rod. It seems to me that the Cromagnard owner of the rod would have wished to see "what the picture really looked like," and so would have on some occasion and more than once have "printed it off" or as we say "unrolled it."

Leaving that question aside I have a few words to say as to the present attempted "completion" of the picture. My difficulty has been in realizing the suggestion of a free, graceful "bounding" action given by the pair of small hind legs which form all that remains of the smallest of the three deer. I have tried various poses of the calf indicated by these legs—bucking and jumping, and with fore legs closely bent to the horizontal or in a more open position. The fact is there is very little in existing drawings or photographs which can help us to a decision of the problem, "How did the prehistoric artist complete that exquisite little pair of hanging legs?" The problem is more obscure even than that of the pose of the arms of the Venus of Melos. One feels sure that the man who made this carving was an artist who must keep a certain rhythm and flow in the action and form of the three successive animals, and it is clear that he was a wonderful observer of the phases of the limbs in movement. It is, perhaps, a presumptuous thing to attempt on such a basis to recall the thought of a man who died twenty thousand years ago, but I set out to do so with the belief that there is a necessary figure determined by those hind legs.

Some years ago, as a step towards a solution of the problem, I published a "restoration" or "completion" of this picture in the "Field" (May 13th, 1911), and asked for criticisms and suggestions from the readers of that journal. I had no difficulty as to the completion of the biggest stag by drawing in his haunches and hind-legs, but the completion of the head and antlers of the smaller stag— and still more the calling into being of the entire calf as an inference from his or her suspended hind-feet and hoofs alone—were not easy tasks. I consulted many authorities and some instantaneous photographs, but I was not satisfied with the pose I finally suggested for the calf nor with the "points" assigned by my draughtsman to the antlers of the smaller stag. Some interesting suggestions were made in reply to my appeal by readers of the "Field." Those which seemed to me of conclusive weight and value were offered by Mr. Walter Winans, who combines the qualifications of a great observer of big game with those of a great artist. In the restoration now given in Fig. 5 I have profited by Mr. Walter Winans' criticism and have been especially glad to make use of the spirited sketch made by him for my benefit, and published in the "Field" of 1911, of a red-deer calf when hopping along with all the feet together, a movement known as "buck-jumping." "Of course," writes Mr. Winans, "this is quite different to the bronco-pony's action when trying to get rid of a rider. In the case of this kind she does not come down with a jar— but as she lands bends her knees and hocks simultaneously and then straightens them, also simultaneously, bounding in the air with bent back, tail curled tight on back, head thrown back, and ears forward; she never puts her fore-legs, either knee or fetlock, beyond her shoulder in this action." These words of Mr. Winans and his outline sketch of the buck-jumping calf precisely realize what the little hanging legs of the rubbed-out calf had been, as it were, urging my tired brain to recall and visualize. I am convinced that Mr. Winans' sketch gives the completion of the picture as drawn by the artist of the Lortet cavern, and satisfies the demand made by the gracefully suspended limbs shown in the incompletely preserved original. And so I have used it in my final restoration here given in Fig. 5.

The following letter by Mr. Winans, giving valuable comments on the Lortet picture, was published in the "Field," and will assist others in appreciating its significance: it enabled me to get the middle stag's antlers correctly drawn. I have omitted a few lines referring to defects in the original restoration—now corrected.

SIR,—As Sir Ray Lankester asks for criticism of this wonderful drawing of three deer, perhaps the following may be of interest. I have known deer all my life, and lived amongst them the last twelve years. I agree that the picture is wonderful—better than anything Landseer or Rosa Bonheur drew, because these latter were only artists: one can see by their pictures (full of faults as to attitudes and actions) that they knew nothing of deer. For instance, Landseer's stags were much too big in the body and their heads too small, and even the shape of their horns was conventional....

"The Lorthet drawings enable one to know all details about the three deer (looking at the original mutilated 'development'). First, the deer have 'got the wind' of an enemy, have come a long way, and are moving leisurely, the big stag, as usual, bringing up the rear and taking a last look round before the herd goes out of sight. The second is the younger stag who generally accompanies the big stag and acts as his sentinel when he is sleeping, a stag too small to give the big stag any jealousy as to his hinds. The third is undoubtedly a calf (Red deer are 'stags,' 'hinds,' and 'calves,' not 'does' and 'fawns'; the latter terms apply to Fallow deer and Roe-deer).

"The deer are typical Red deer, not Wapiti, except that the only tail showing (that of the middle deer) is the short Wapiti tail, not the longer tail of the Red deer, and the ears are shorter than those of any existing species of deer.

"The horns of the big stag are those of typical park Red deer, exactly like the Warnham Park big stag: brow, bay, and tray, with a bunch on top, and the horns are short and straight for their thickness.

"Now as to the short tail. I am trying, by crossing the Wapiti, Red deer, and Altai to get back to the original deer before the various species got separated, and my 'three-cross' deer show these very characteristics, as follows: Red deer or Warnham horns, short Wapiti tail, and the rather Roman nose which this 'development' print shows. The only difference is the short ears. Is it not possible that, as the artist is able to draw the horns in perspective and show the anatomy and proportions so well, that the ears are meant to be drawn fore-shortened?

"The stag's mouth is open because he is big and fat and is blowing (not roaring or bellowing). If it was the rutting season, when stags roar, the stag would be tucked up in the belly and have a tuft of hair hanging under the middle of it. He and the stag in front are moving in the real action (not the conventional action Rosa Bonheur and Landseer drew, but what the ancient Egyptians drew sometimes) of a slow, easy canter.... Now as to the middle stag's horns. I should give him, bearing in mind he is the small sentry stag, brow, tray, and three on top—a ten-pointer, the thin points showing in the original drawing indicating that he had thin horns—in fact, a three-year old.

"In a Scotch forest a ten-pointer is a comparatively old stag, but at Warnham and my place, where the feeding is good (and in my case there is hand feeding all the year round), a spike stag gets six points and can almost be a royal the next year.

"All this shows that the deer at the time this drawing was made must have had very good feeding and come to maturity quickly, like modern park deer. The big stag would never have allowed a ten-pointer in his herd if the latter had been an old stag.

"As to the action of the leading hind. I think she is a hind-calf by her legs, and is jumping with all four legs together, the way young deer do when playing, and, being young, is paying no attention to the danger behind, but is full of life, like a horse playing about when he is fresh. One often sees the calves of a herd playing like this if the herd is moving along steadily....

"From the position of the hind legs of the little calf I judge that she is jumping with all four legs together (the jump from which the expression 'buck jumping' comes); her tail would be curled up tight over her back like a pug dog carries it, only without the curl, and her ears pricked forward. The piece of horn broken off would show the rest of the hinds and calves, led by an old 'yeld' (*i.e.*, barren) hind, who would be leading the herd up wind with her nose and ears forward to 'get the wind' of any danger ahead.

"The day is a hot one in the middle of August, shown by the big stag blowing and his being with the hinds, instead of with other stags by themselves, and by his not having 'run' yet, though his horns are clear of velvet. He is most likely the stag on whose horn this is engraved. The length of the deer's feet shows that they live on ground which is soft and not many stones about to wear down their toes.

"Maybe the fish indicate that the deer are crossing a shallow ford, and the salmon are getting frightened and jumping. The right-hand-most fish is just in the attitude of a hooked salmon trying to leap clear of the fly....

"The picture was most likely first drawn on some flat flexible surface, skin or bark, in a sticky medium, and then transferred to the horn by rolling it round the horn and then rubbing it. This would give a transfer, which would guide the subsequent engraving, otherwise it would be very difficult to engrave direct on the horn, and mistakes could not easily be corrected.

"WALTER WINANS

"SURRENDEN PARK, PLUCKLEY, KENT

With regard to the six fishes in the picture of "The Three Red Deer," I think that there can be little doubt that they are put in in the same spirit of exuberance which induced early Italian masters to introduce a cherub wherever a space for him could be found. The fish represented are the same in each case, and are undeniably salmonids. Presumably they are drawn on a larger scale than the deer. Their markings and the form of the head are deserving of some criticism and comment by those who are familiar with fish as seen by the fisherman. Probably the artist's friends at Lourdes captured fish in those days by spearing them with serrated bone-headed fish spears or harpoons (Fig. 3). No fish hooks of bone have been found in the cave of Lortet or in others of like age, although needles and whistles of bone and other useful little instruments, as well as serrated spear heads and harpoons have been obtained in several of them.

The tool used by the prehistoric man in engraving the cylinder of stag's antler was undoubtedly a suitable chipped-out piece of flint—a flint graving tool, in fact a "burin," such as are abundant in these caves.

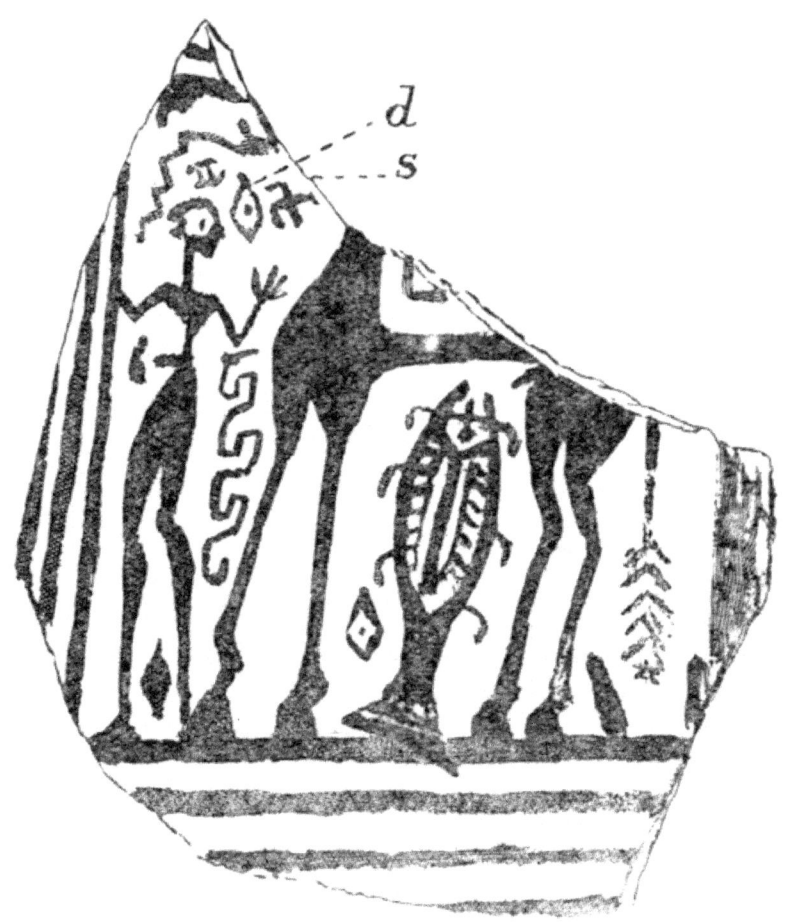

FIG. 6.—Fragment of a roughly-painted vase of the Dipylon age (*circa* 800 B.C.) from Tiryns, figured by Schliemann and cited by Hörnes in his "History of Pictorial Art in Europe." Compare the fish between the horse's legs with the fish in the Lortet picture of the Three Deer; also note the lozenge-shaped designs (similar to the pair above the big stag in the Lortet picture) near

the fish and near the man's head (*d*); and, further, the swastika (*s*).

Attention has been drawn by Hörnes in his "History of Pictorial Art in Europe" to the resemblance of the Lortet picture to a fragment of a roughly painted vase of the Dipylon age (*circa* 800 B.C.) found at Tiryns and figured by Schliemann in his account of excavations made at that ancient Mykenæan fortress of the Peloponese. The fragment (Fig. 6) shows very roughly drawn figures of a man and of a horse. Between the fore and hind legs of the horse a large elaborately ornate fish is represented, reminding us of the fishes between the deer's legs in the Lortet picture. Two other similar fragments of pottery, showing a fish in this position, are recorded by Schliemann. The drawing is conventional and careless. It is of a debased decorative character, and is very far removed from the careful nature-true work of the Lortet cave-man. It is not possible to trace by any known line of transmission a connection between the engraving executed 20,000 years ago in the caves of the Pyrénées and the figures rapidly knocked off in black paint on the Tiryns vase some 17,000 years later by the local dealers in cheap pottery. Yet we cannot avoid the suggestion that there is some connection between the two designs. For the Tiryns painting shows not only the curious upright fish between the horse's legs, but also diamond-shaped figures—one marked *d* in Fig. 6, another near the fish's tail, and another between the man's feet—closely resembling the pair of diamond-shaped figures engraved above the neck of the big stag in the Lortet picture (see Figs. 4 and 5). As we do not know what these diamond-shaped figures or "lozenges" are intended to signify in either case, we do not get, at present, beyond the bald fact of their coincidence. The Tiryns painting also shows (at *s* in Fig. 6) a "swastika" (see Chapter XVII), and below the man's arm a carelessly drawn bit of the ancient wave-fret or key-pattern. It is, of course, possible that the tradition of an ancient design— even dating so far back in origin as many thousands of years—may be preserved in the use made in the Tiryns decoration of the fish and the diamond-shaped lozenges, though associated with the swastika and the bit of wave-fret which are probably of later origin and are not known in the decorative work of the cave-men. The Mykenæan decorative assimilation of geese to the ship's barnacle exercised its influence over three thousand years and led to the mediæval belief in the hatching of young geese from barnacles attached to floating timber, and even from the buds of trees (see my "Diversions of a Naturalist": Methuen, 1915). Nevertheless it must not be supposed that the connection of the Lortet engraving and the vase-painting of Tiryns is probable or more than a very remote possibility. The gap in time is too vast, and our present ignorance of what took place in that interval too complete, to warrant us in regarding the resemblance as more than a coincidence.

FOOTNOTES:

[1] "L'Age du Renne," a posthumous work, with one hundred coloured quarto plates of objects in the Piette collection, is published by Masson, of Paris, and gives the complete list of Piette's numerous earlier papers, issued as his excavations proceeded.

[2] Seven years ago the ape-like lower jaw and thick walled brain-case called "Eoanthropus" were discovered in a sparse gravel near Lewes in Sussex. It is probably of older date than either the Neander men or the Heidelberg men. See on this subject the chapters on "The Missing Link" in my "Diversions of a Naturalist" (1915) and those on "The Most Ancient Men" and "The Cave-men's Skulls" in "Science from an Easy Chair. First Series" (1910).

CHAPTER II

PORTRAITS OF MAMMOTHS BY MEN WHO SAW THEM

SOME fifty-five years ago pieces of reindeer's antler were discovered in the cave known as "La Madeleine" in the Dordogne (a department of France some eighty miles east of Bordeaux), upon which were engraved the outlines of various animals such as reindeer and horses. They and the bone spear-heads and needles, and the flint knives found with them, were the first revelation to later man of the existence of the prehistoric cave-men. Among the carvings was a piece of ivory which excited the profoundest interest. Partly hidden by a confused mass of scratches it showed the well-drawn outline of the great extinct elephant, thus scratched or "engraved" on a bit of its own tusk (Fig. 7). The engraving was barely 5 in. long, and has been reproduced in many books. The specimen is now in Paris, and was for long the only known representation of the Mammoth by the ancient men who lived with it in Western Europe.

FIG. 7.—Engraving of a mammoth drawn upon a piece of mammoth's ivory, found in the cave of La Madeleine in the Dordogne, in 1864. The specimen is in the Museum of Natural

History, Paris. The engraving is here represented of the actual size.

During the last fifteen years, however, our knowledge of the works of art executed by these ancient men has increased to an extraordinary extent, chiefly owing to the energy and skill of the French explorers of the caverns in the south central region of that country. As long ago as 1879 a little girl, the daughter of Señor Sautuolo—a proud woman she should be if alive to-day—when visiting the cavern of Altamira, near Santander, in the north of

Spain, with her father, drew his attention to a number of "pictures of animals," painted on the rocky vault or roof of the cave. At first no one believed that these pictures were more than a few hundred years old, whilst some held them to be modern and made with fraudulent purpose. In 1887 Piette, the distinguished French investigator of the remains of human work in the caverns of the French Pyrénées (whose great illustrated book of carved and engraved portions of reindeer antler, ivory, and stones discovered by his excavations, is a classic), declared that in his opinion the pictures of the Altamira cave were of the same age as the bone and ivory carvings of the Madeleine cave—that is to say, dated from what "prehistorians" call the later Palæolithic age, an age when the mammoth, the bison, the cave lion, and the reindeer still existed in Western Europe, and when the British Isles were not yet separated by sea from the Continent. The age indicated is probably from 25,000 to 50,000 years ago. Still, the opinion prevailed that the "wall-drawings" and "roof-drawing" of the Altamira cave were either mediæval or modern until the French explorers discovered wall-paintings in some of the caves of the Dordogne. Then they proceeded to a careful investigation of the Altamira cave, and discovered conclusive evidence of the great age of the paintings by the removal of some of the undisturbed deposit in the cave, in which were found flint implements and small engravings on bone, proving the deposit to be of the late Palæolithic age. When this deposit was removed, pictures of animals, partly engraved and partly completed in colour (black, red, yellow, and white), were found on the wall of the cave previously covered up by the deposit. M. Cartailhac, who had been a leading opponent of the view that the Altamira wall-pictures were very ancient, now renounced his former position and became an enthusiastic investigator and exponent of these pictures. M. Breuil, who had discovered wall-pictures, including those of the mammoth, in French caves, and had been met by disbelief and even suspicion, now received due recognition, and joined Cartailhac in preparing a complete account of the wall and roof pictures of the Altamira cave. The Prince of Monaco, who had carried out, with the aid of French experts, an investigation of the caves on his property at Mentone, on the Mediterranean "Riviera," undertook the expense of producing a splendid volume, giving coloured reproductions of the Altamira pictures. To him the world is indebted, not only for most important discoveries of human skeletons and objects of human workmanship in the caves of Mentone (there are no wall-pictures there), but for the publication in illustrated form of the Mentone discoveries and of those obtained in the Altamira cave. He has not rested at this stage of accomplishment, but has produced at his own expense large volumes by MM. Breuil, Capitan, and Peyrony, illustrating and describing the discoveries made by them of wall-paintings and engravings of animals in the cave known as the "Font de Gaume," in the Dordogne. The Prince has also published a volume, by MM. Breuil, de Rio, and Sierra, reproducing the drawings found in a whole series of caves and rock-shelters in various parts of the Spanish peninsula, where the rock-painting race seems to have persisted to a somewhat later period and to have painted, more frequently, pictures of human beings as well as of animals. These, whilst less artistic and truthful than those of the North Spanish and South French area, yet have surpassing interest, since they have special similarity to ancient rock-paintings found in North Africa and to the rock-paintings of the Bushmen of South Africa.

The Prince of Monaco has finally established the great study in which he has played so valuable a part by founding in Paris an "Institute of Human Palæontology"; that is, "of the study of prehistoric man," which he has endowed with a magnificent building, comprising laboratories and residences for professors, together with funds to pay for its maintenance and the proper publication of results. This he has done in addition to founding entirely at his own expense a similarly complete Institute for the study of "oceanography"—the study of the living contents and history of the great seas.

The illustrations in this chapter are (with the exception of Fig. 7) copies, greatly reduced in size, of faithful representations of the great hairy elephant or mammoth which still survived in southern France in the days when the caves were occupied and decorated by men. I am indebted to the valuable little book "Repertoire de l'Art Quatermaire," by M. Salomon Reinach, for these outlines carefully drawn by him from various large illustrations by the use of a tracing and reducing instrument. In the next chapter I have given examples from the same source of similar drawings of other animals.

There are five kinds of artistic work of Palæolithic age found in the caverns of France and Spain; namely (1) small solid carvings (complete all round) in bone, ivory, or stone; (2) small engravings in sunk outline on similar material, rarely with relief of the outlined figure; (3) large stone statues, 2 ft. to 6 ft. across, in high relief, with complete modelling of the visible surface; (4) rock engravings and paintings on the walls and roofs of caverns or rock shelters, often partly outlined by engraving and scraping of the surface, and then completed in black or red paint or in several colours (black, red, yellow, white); they are of large size, from 2 to 5 ft. in cross measurement; (5) models in clay, one side only shown, the other resting on rock; a few incomplete clay models of this nature representing the bison of about 2 ft. in

length, have recently been discovered in one of the French caverns, and are the only examples of modelling in clay by the Palæolithic men yet discovered.

FIG. 8.—Outline engravings of mammoths on the wall of the cavern
known as the "Font de Gaume," near Eyzies (Dordogne). Each
figure is about 2 ft. long.

Our figures of the mammoth are (excepting Fig. 7) all of
the fourth class—namely, rock-paintings in one colour
(black or red) partly engraved and scraped. The originals
are from 1-1/2 ft. to 2-1/2 ft. long. The mammoths given in
Fig. 8 are carefully copied from engravings discovered,
reproduced, and described by M. Breuil and his fellow-
workers. They are on the walls of the cavern known as the
"Font de Gaume," in the commune of Tayac in the
Dordogne. Those copied in Fig. 9 and Fig. 10, A, were

discovered on the walls of the cave of Les Combarelles in the same district.

FIG. 9.—Similar engravings from the neighbouring cave of Combarelles. The lower figure is an enlargement of the smaller of the two above it.

Fig. 10, B, is from a cave at Bernifal, near les Eyzies, in the Dordogne, and shows a mammoth enclosed in a triangular design, which is believed to represent a trap, or else a cage. Such triangular figures with upright and also bent supports are found in various degrees of elaboration on both small and large engravings of this period, and are generally accepted as representing huts or enclosures supported by wooden poles. They are called "tectiforms" by the French explorers.

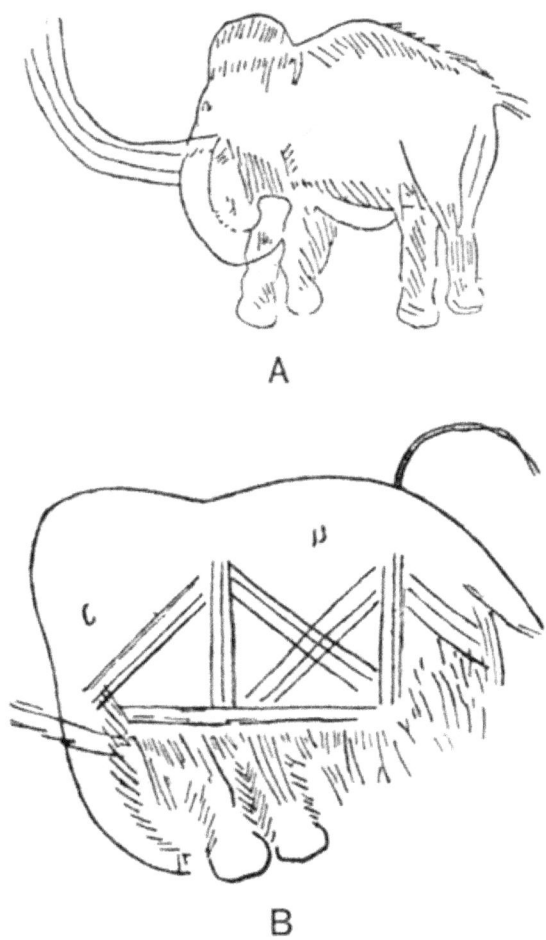

FIG. 10.—*A*, similar engraving from the cave of Combarelles. *B*, Mammoth enclosed by plank-like structure—supposed to be either a cage or a trap. (Called tectiform structures, and often seen in these wall engravings.) From the cave of Bernifal, five miles from Eyzies.

The bones and teeth of the mammoth are very common in the river gravels and clays of Western Europe and England, and a complete skull, with its tusks, dug up at Ilford, in the east of London, is in the Natural History

Museum. Frozen carcasses of this animal are found in Northern Siberia, and two showing much of the skin and hair are in the museum of Petrograd. There is no tradition or knowledge of the mammoth among living races of men. The natives of Siberia, who have from time immemorial done a large trade in the ivory, regard the tusks as "horns," and have stories about the ghosts of the mammoth, but no tradition of it as a living beast. The mammoth was closer to the Indian elephant of to-day than to the African one. It had, as these drawings show, a pelt of long hair. Indian elephants from upland regions often have a good deal of hair all over the body: and the newborn young of both the Indian and African elephant has a complete coat of hair. The drawings here reproduced are not only of thrilling interest because they are the work of remotely ancient men who lived with and observed mammoths in the south of France, but also because they show an extraordinary skill in "sketching"—in giving the essential lines of the creature portrayed and in reproducing the artist's "impression." These artists were "impressionists"—the earliest and most sincere—without self-consciousness or other purpose than that of making line and colour truly register and indicate their vivid impressions. It is interesting to note that (as in other works of art showing true artistic gift) actual error in drawing (for instance, in the size and shape of the eye and the placing of the two tusks on the same side of the trunk—possibly due to the unfinished state of the drawing) sometimes accompanies the most penetrating observation and skilful delineation of the characteristic form and pose of the animal. Probably mammoths were getting rare in the south of France when these drawings were made, and were not so familiar in all their details to the artist as were bison, horse, and deer.

CHAPTER III

THE ART OF PREHISTORIC MEN

THE works of art produced by the cave-men are, as we have already seen, of five kinds or classes—(1) All-round small statuettes, or "high-relief" carvings, in ivory, bone, or stone (examples of which are shown in Figs. 14, 25, 26, 27, 28 of the present chapter); (2) small engravings on bits of ivory, deer's antler, bone, or stone (examples are shown in Figs. 15, 16, 20, and 24); (3) large statues, hewn in rock, and left in place; (4) drawings of large size—two to five feet in diameter (partly engraved and partly coloured) on the rocky walls and vaults of limestone caverns (shown in Figs. 11, 12, 13, 17, 18, 19, 23, as well as in the figures of mammoths in the last chapter); (5) models (high relief) worked in clay. I give reproductions in the present chapter of several samples of this art, showing how skilfully these men of 50,000 years ago could portray a variety of animals.

Who were these men, and why did they make these remarkable carvings and drawings? First, as to their age. We now know of a long succession of human inhabitants of this part of the world, namely, Western Europe. The earliest reach back to an antiquity never dreamed of fifty years ago. We cannot fix with any certainty the number of thousands, or hundreds of thousands, of years which is represented by this succession, but we can place the different periods in order, one later than the other, each distinguished chiefly by the character of the workmanship belonging to it, though in a few instances we have also the actual limb-bones, skulls, and jaw-bones of the men themselves, which differ in different periods. It is practically certain that these prehistoric successive periods of humanity do not represent the steps of growth and change of one single race belonging to this part of the world, but that successive races have arrived on the scene of Western Europe from other parts, and it is usually very difficult even to guess where they came from and where they went to!

It is convenient to divide the human epoch, the time which has elapsed since man definitely took shape as man—characterized by his large brain, small teeth, upright carriage, and large opposable thumb and still larger and more peculiar non-opposable great toe—into the historic and the prehistoric sections. In this part of the world (Europe) the first use of metals (first of all copper, then bronze, and then iron), as the material for the fabrication of implements and tools of all kinds, occurs just on the line between the historic and the prehistoric sections; that is to say, between those times of which we know something by tradition and writing, and those earlier times of which we have no record and no tradition, but concerning which we have to make out what we can by searching the refuse heaps and ruins of man's dwelling-places and carefully collecting such of his "works" as have not utterly perished, whilst noting which lie deeper in the ground, which above and which below the others.

Practically the men of the prehistoric ages in Europe had not the use of metals (though our quasi-historical records go back to a less remote time in many parts of Europe than they do in Greece, Assyria, and Egypt). The prehistoric peoples are spoken of as the men of the Stone Age, because they used stone, chiefly flint, as many savage races do to-day, as the material from which they fabricated by means of deftly struck blows all sorts of implements. Undoubtedly they also, by aid of stone knives, saws and planes, made weapons and other implements of wood and of the horns, bones, and teeth of animals. But these latter substances are perishable, and have only been preserved from decay under special circumstances, such as their inclusion in the deposits on the floors of caverns.

The Stone Age is itself readily and obviously divisible into two periods. The latter is a comparatively very short and recent period, when great skill in chipping flints and other stones was attained, and the implements so shaped were often rubbed on large stones of very hard material (siliceous grit), so as to polish their surfaces. This is the "Neolithic," or later Stone, period, and extends back in Europe certainly to 7000 B.C., and probably a few thousand years further. Passing further back than this, we leave what are called "recent" deposits, and come to those associated with great changes of the earth's surface. We enter upon "geological" time, and vastly changed climatic and geographical conditions. We are in the older Stone period, called the "Palæolithic period." It is not really comparable to the "Neolithic," since it comprises many successive ages of man, and, although called the "Palæolithic" or "ancient Stone" period, has no unity, but, whilst readily divisible into several sub-periods or epochs of comparatively late date, stretches back into immense geologic antiquity indicated by flint implements of special and diverse types, which are found in definitely ascertained geologic horizons.

The Pleistocene strata—the latest of the geologists' list—are the river gravels of existing river valleys, the deposits in many caves, and the sands and clays piled up by ice action during the repeated glacial extensions or epochs of glaciation which alternated with milder climate for many thousands of years over north and middle Europe. It is identical with the Palæolithic period, which, however, probably extends beyond it into the Pliocene and even further back. In the later deposits of the Pleistocene, which necessarily have been less frequently disturbed and re-deposited than the older ones, we find more numerous remains of man's handwork, and in less disturbed order of succession, than in the older deposits. Lately we have obtained in East Anglia beautifully-worked flint implements—the rostro-carinate, or eagle's beaks—from below shelly marine deposits—the Red Crag of Suffolk and the Norwich Crag—the oldest beds of the Pleistocene. They were made by men who *lived* in the Pliocene period, and carry the ancient Stone period of man back to a much earlier period than was admitted nine years ago.

The Pleistocene series or "system " of strata—also called the "Quaternary" to mark its distinction from the underlying long series of "Tertiary" strata—does not comprise the actual surface-deposits in which the remains of Neolithic man are found. It is usual, though perhaps not altogether logical, to separate these as "Recent" and to begin the long enumeration of "geologic" strata after a certain interval when the relative levels of land and sea and the depth of river-valleys were not precisely what they are to-day, and the human inhabitants of Western Europe were hunters using rough unpolished flint implements—in fact, when the "Palæolithic" period of human culture had not given place to the "Neolithic," which was after some ten thousand years itself to be superseded by the age of metals. "Prehistorians," the students of prehistoric man—divide the Pleistocene series of deposits with a view to a systematic conception of the successive changes of man and his surroundings during the period occupied by their deposition, into an upper, a middle and a lower group— and further have distinguished certain successive "horizons" in these groups—characterized by the remains of man and animals which they contain. They are exhibited in the tabular statement here given in the ascertained order of their succession, and are represented in the southern part of Britain as well as in France.

HORIZONS OR EPOCHS OF THE PLEISTOCENE OR QUATERNARY SYSTEM

A. UPPER PLEISTOCENE (post-glacial; also called epoch of the Reindeer).

1. *The Azilian:* (Elapho-Tarandian of Piette) nearest to the Neolithic section of the Recent Period and more or less transitional to that period; named after the cavern of the Mas d'Azil in the department of the Ariège. The Reindeer had largely given place to the great Red Deer (Cervus elephus).

2. *The Magdelenian:* named after the cave of La Madeleine in the Dordogne.

3. *The Solutrian:* after Solutré near Macon.

4. *The Aurignacian:* after the grotto of Aurignac in the Haute Garonne.

B. MIDDLE PLEISTOCENE (period of the last great extension of glaciers).

1. *The Moustierian:* so named after the cave of Le Moustier in Dordogne; the epoch of the Neander men. Also called the "epoch of the Mammoth," whilst the upper Pleistocene is called the epoch of the Reindeer, though the Mammoth still survived then in reduced numbers.

C. LOWER PLEISTOCENE (inter-glacial and early glacial, also called period of the Hippopotamus and of Elephas antiquus and Rhinoceros Merckii).

1. *The Chellian:* named after Chelles on the upper Seine, river gravels and sands earlier than the Moustierian. Large tongue-shaped flint implements, flaked on both surfaces—the later and better-finished classed as "Acheulæan," after St. Acheul, near Amiens.

2, 3, 4... various fluviatile and lacustrine gravels, sands and clays divisible into separate successive horizons, as well as marine deposits, some of glacial origin—including the mid-glacial gravel, the boulder clays and shelly Red Crag and Norwich Crag (but *not* the underlying "Coralline" Crag, which must be classed with the Pliocene). The relations of the marine deposits to the older river-gravels and fresh-water deposits, and to the earlier periods of glacial extension indicated by the glacial moraines of central Europe, have not been, as yet, satisfactorily determined.

The amount of the sedimentary deposits of the earth's crust belonging to the Pleistocene or Quaternary Period—about 250 feet in thickness—is exceedingly small, and represents a surprisingly short space of time as compared with that indicated by the vast thickness of underlying deposits. It has nevertheless been possible to study and classify the "horizons" of this latest very short period minutely because the deposits are easily excavated, and having been more recently "laid down" have not suffered so much subsequent breaking up and destruction as have the older strata; and further, because they embed at certain levels and in favourable situations an abundance of well-preserved bones and teeth of animals and the implements and carvings in stone and bone made by man. It is worth while to look at this matter a little more exactly.

The total thickness of sedimentary deposits—that is, deposit laid down by the action of water on the earth's surface, and now estimated by the measurement of strata lying one over the other in various parts of the globe—tilted and exposed to view so that we can trace out their order of super-position—is about 130,000 feet. The lower half of this huge deposit contains no fossilized remains of the living things which were present in the waters which laid it down; they were soft, probably shell-less and boneless, and so no fossilized trace of them is preserved. Thus we divide the sedimentary crust into 65,000 feet of "archaic" non-fossiliferous deposit, and an overlying 65,000 feet of fossil-containing deposits.

The earliest remains of living things known are not very different from marine creatures of to-day; they are the strange shrimp-like Trilobites and the Lingula-shells found in the lower Cambrian rocks of Wales. Over them lie 65,000 feet of sedimentary deposit teaming with fossils— the petrified remains of animals and plants. The Trilobites and the Lingulas must have had a long series of ancestors leading up to them from the simplest beginnings of life— for they are highly organized creatures. But no trace of those ancestors is preserved in the 65,000 feet of sedimentary rock underlying the earliest fossils.

This great basal mass of non-fossiliferous deposit is called "the Archæan series." The 65,000 feet of deposit *above* it are divided by geologists into three very unequal series. The first and lowest is the Primary or Palæozoic series, occupying the enormous thickness of 52,000 feet; above these we have the Secondary or Mesozoic series of 10,000 feet, and lastly, bringing us to recent time, we have the Tertiary or Cainozoic of only 3000 feet. These three series amount in all to 65,000 feet. The Palæozoic series is more than five times as thick as the Mesozoic, and these two taken together are twenty times the thickness of the Tertiary. Each series is divided by geologists into a series of systems, distinguished by the fossils they contain, which, on the whole, indicate animals of a higher degree of evolution as we ascend the series.

The Palæozoic series include the vast thicknesses of the Cambrian, the Ordovician, the Silurian, Devonian, Carboniferous and Permian systems. The first "trilobite" is found in the lowest Cambrian rocks, and the last or most recent existed in the Permian period—after 50,000 feet of rock had been deposited. None are known of later age. The first fossil remains of a vertebrate are found in the uppermost beds of the Silurian—in "beds" (that is to say, stratified rocks) which are just *half-way* in position so far as the measurable thickness of the deposits are concerned, between the earliest Cambrian fossils and the sediments of the present day. To put it another way, 34,000 feet of fossiliferous rock precede the stratum (upper Silurian) in which the earliest remains of vertebrates are found. These first vertebrates to appear (others soft and destructible preceded them) are fishes—a group which, apart from this fact, are shown by their structure to present the ancestral form of all the vertebrate classes. In later Palæozoic beds we find the remains of four-legged creatures like our living newts and salamanders. The Secondary or Mesozoic series is divided into the Triassic, Jurassic and Cretaceous systems. It ends with the familiar chalk deposit of this part of the world, and is often called the age of Reptiles, because large reptiles abounded in this period. The Tertiary or Cainozoic series are divided into the Eocene, Oligocene, Miocene, Pliocene and Pleistocene systems. The huge reptiles disappear and their place is taken by an endless variety of warm-blooded, hairy animals—the Mammals—small at first, but in later beds often of great size. As we pass upwards from the Eocene we can trace the ancestry of our living Mammals such as the horse, rhinoceros, pig and elephant in successive forms. Complete skeletons are preserved in the rocks and show a gradual transition from the more primitive Eocene kinds—through Miocene and Pliocene modifications—until in the Pleistocene strata many of the species now inhabiting the earth's surface are found. A number of horizons, characterized by the special mammalian and other animal remains preserved in them, are distinguished by geologists in each of the "systems" of sands, clays and harder beds known as Eocene, Oligocene, Miocene and Pliocene. At last we arrive at the latest or most recent 250 feet of deposit, consisting of sand, clay and gravel. This is called "Pleistocene." It is only a very small fraction (1/260th) of the thickness of the whole fossil-bearing sedimentary crust of the earth—about the proportion of the thickness of a common paving-stone to the whole height of Shakespeare's cliff at Dover. This Pleistocene or post-glacial Tertiary—often now called Quaternary—has been so carefully examined that we divide it as shown on page 39 into upper, middle and lower, and each of these divisions into successive horizons (only a few feet thick) characterized by the remains of different species of animals and often by the differing implements and carvings as well as the bones of successive races of men.

When we are concerned with written history, ancient Egypt seems to be of vast and almost appalling antiquity; on the other hand, if we study the cave-men, ancient Egypt becomes relatively modern, and the first cold period and extension of glaciers, which 500,000 years ago marked the passage from Pliocene to Pleistocene, becomes our familiar example of something belonging to the remote past—beyond or below which we rarely let our thoughts wander. That is a natural result of concentration on a special study. But it has had the curious result, in many cases, of making students of ancient man unwilling to admit the discovery of evidences of the existence of man at an earlier date than that which belongs to the deposits and remains to which their life-long studies have been confined and upon which their thought is concentrated. The last 500,000 years of the earth's vicissitudes, which resulted in the 250 feet of "Pleistocene" deposit and the marvellous treasures of early humanity embedded in them, form but a trivial postscript to the great geological record which precedes it.

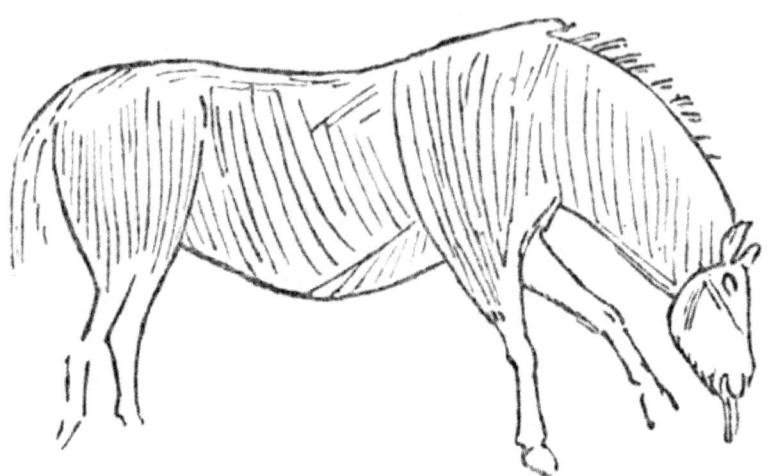

FIG. 11.—Horse (wall engraving), cave of Marsoulas, Haute Garonne. The drawing suggests the Southern less heavy breed as compared with Figs. 12 and 13.

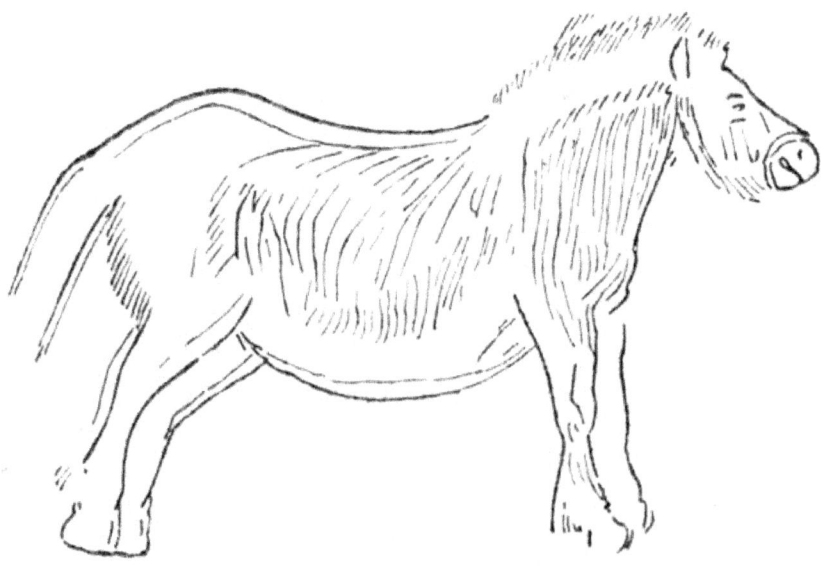

FIG. 12.—Horse (wall engraving) outlined in black, cave of Niaux (Ariège).

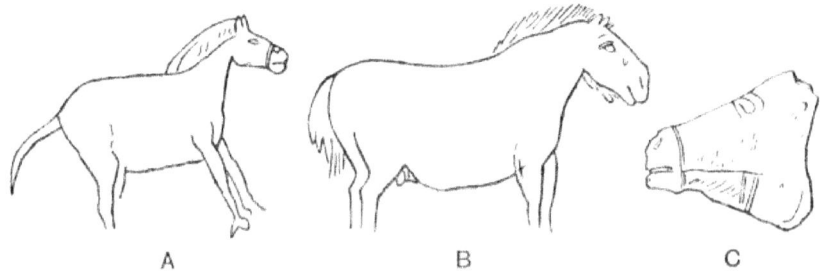

FIG. 13.—Horses: *A*, wall engraving (cave of Hornos de la Péna). *B*, wall engraving from cavern of Combarelles. *C*, engraved on reindeer antler (Mas d'Azil). Note the halter in *A* and in *C*; also note the heavy head and face of *B* like that of Prejalvski's horse.

No estimate can be made of the time represented by the 65,000 feet of fossiliferous strata known to us and the same thickness of non-fossiliferous deposit which precedes them. There are no facts known upon which a calculation of the related lapse of time can be based. But most geologists would agree that whilst we have good ground for assigning half a million years to the formation of the Pleistocene strata, it is not an unreasonable supposition that the period required for the formation of the fossiliferous rocks which precede them in time, is not less and probably more than five hundred million years.

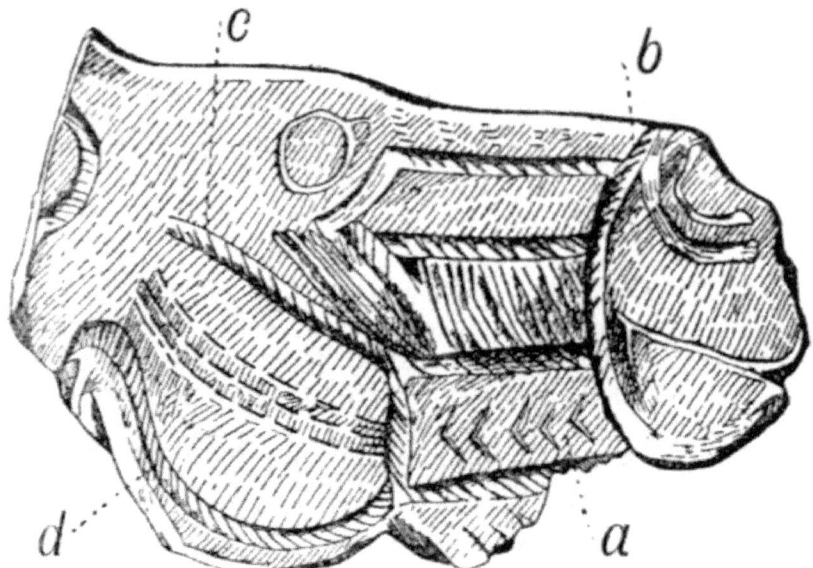

FIG. 14.—Drawing (of the actual size of the original) of a flat carving in shoulder-bone of a horse's head, showing twisted rope-bridle and trappings. *a* appears to represent a flat ornamented band of wood or skin connecting the muzzling rope *b* with other pieces *c* and *d*. This specimen is from the cave of St. Michel d'Arudy, and is of the Reindeer period. This, and others like it are in the same museum of St. Germain.

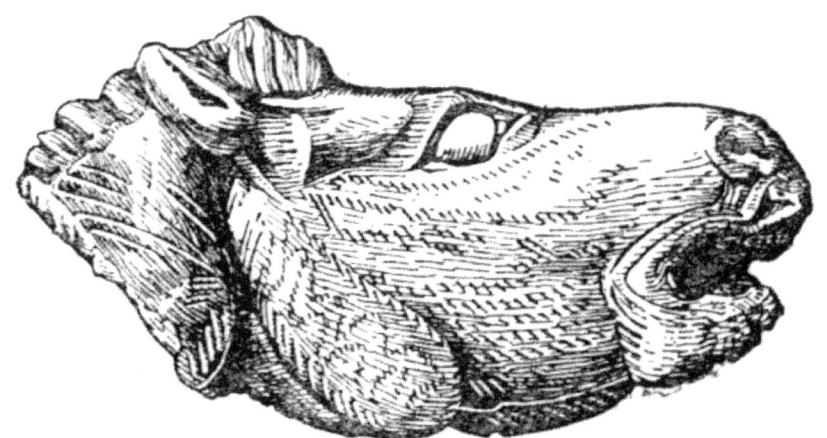

FIG. 15.—Drawing (of the actual size of the original) of a fully rounded carving in reindeer's antler of the head of a neighing horse. The head resembles that of the Mongolian horse. This is one of the most artistic of the cave-men's carvings yet discovered. It is of the Palæolithic age (early Reindeer period), probably not less than 50,000 years old. It was found in the cavern of Mas d'Azil, Ariège, France, and is now in the museum of St. Germain.

The pictures and carvings with which we are for the moment concerned all belong to the *later* Pleistocene or Reindeer epoch. None have been found in the middle and earlier Pleistocene, though finely-chipped flints of several successive types are found in those earlier beds. So that it is clear that many successive ages of man had elapsed in Western Europe before these pictures—immensely ancient as they are—were executed. The men who made these works of art had ages of humanity, tradition, and culture (of a kind) behind them. Yet they were themselves tens of thousands of years earlier than the ancient Egyptians!

FIG. 16.—Reindeer engraving on schist, small size (cavern of Laugerie basse).

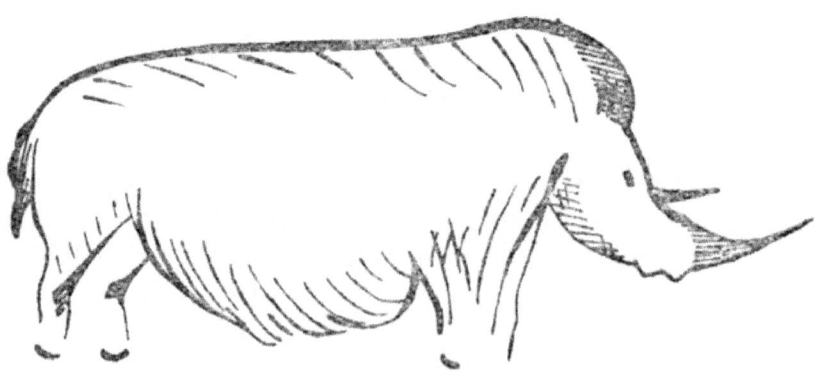

FIG. 17.—Rhinoceros in red outline (2-1/2 feet long), drawn on the wall of the cavern of Font de Gaume.

Our illustrations show a variety of drawings and carvings. It appears probable that the primitive intention of ancient man in depicting animals was "to work magic" on those which he hunted. This is the case at the present day among many "savage" races. The drawings of bisons in Fig. 19 are from the walls of the cavern of Font de Gaume, in the Dordogne, and are about 5 ft. long, partly engraved and scraped, partly outlined in black, and coloured. The body is often coloured in red, white and black, so as to give a true representation of the masses of hair and surface contours. A specially well preserved painting of this kind—from the cavern of Altamira—is shown in Fig. 18, where the colours of the original—black, red, and brown, and white are indicated by the varied shading. These drawings, like those of the mammoths figured in the last chapter, are found in the recesses of caverns where no daylight reaches them, and must have been executed and viewed by aid of torch or lamp-light. They probably were exhibited as part of a ceremony connected with witchcraft and magic. These, like the mammoths and all the specimens figured here, were executed in the Reindeer, or later Pleistocene period. The exact "horizon" of each is, as a rule, well ascertained, but there is uncertainty as to whether some specimens should be attributed to the Aurignacian or to the Magdalenian horizon—and as to whether work by men of the Magdalenian race is not in some cases associated in the cave deposits with that by the earlier negroid Aurignacians.

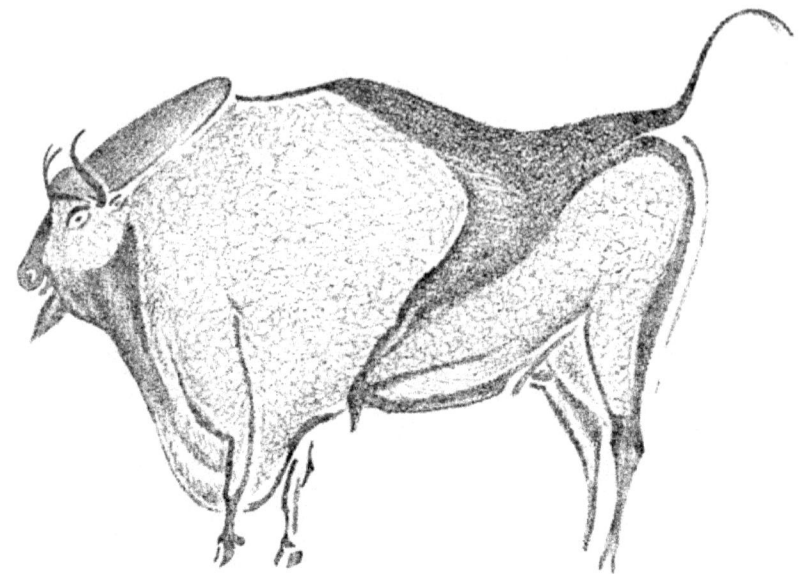

FIG. 18.—Bison from the roof of the cavern of Altamira: engraved, and also painted in three colours (5 feet long).

FIG. 19.—Bison: wall engravings (5 feet long) filled in with colour (Font de Gaume).

FIG. 20.—Bear: engraved on stalagmite, from the cave of Teyjat near Eyzies. (Small size.)

FIG. 21.—Bear, engraved on stone, Massol (Ariège).

FIG. 22.—Wolf, engraved on wall of the cave of Combarelles.

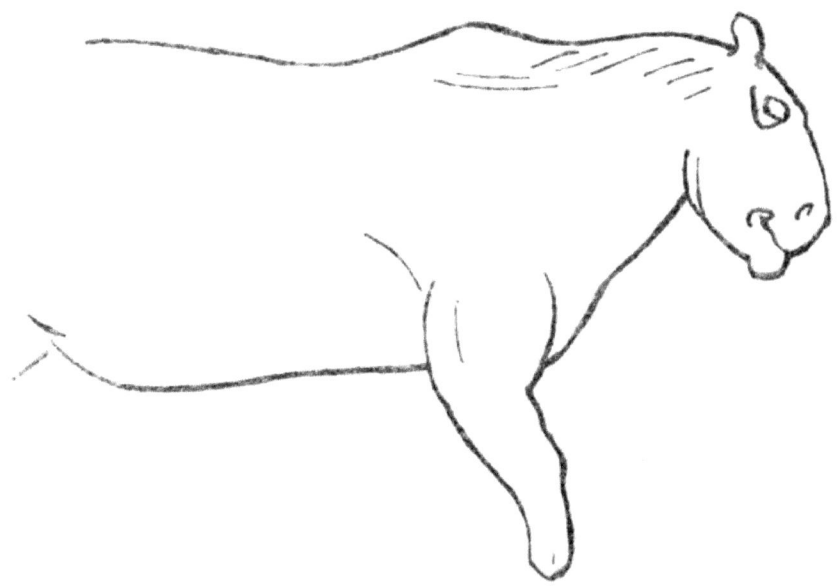

FIG. 23.—Wall engraving of a Cave Lion (Combarelles).

The horses shown are from various caves. Fig. 12 is drawn in black on the wall of a cave at Niaux (Ariège), and Fig. 11 is a similar drawing from a cave in the Haute Garonne. Both are remarkable for the exact representation of natural poses of the horse. Figs. 13, A and B, are also from the walls of caves. The latter is remarkable for the large head, short mane, and thick muzzle, which closely correspond with the same parts in the existing wild horse of the Gobi desert in Tartary (to be seen alive in the Zoological Gardens in London). The horse drawn in Fig. 11 seems to belong to a distinct race, suggesting the Southern "Arab" horse rather than the heavier and more clumsy horse of the Gobi desert. Fig. 13, C, is engraved of the size here given, on a piece of reindeer's antler. It is remarkable for the halter-like ring around the muzzle. A similar cord or rope is seen in Fig. 12 and in Fig. 13, A.

FIG. 24.—Goose: small engraving on reindeer antler (Gourdan).

The most remarkable horses' heads obtained are those drawn (of the actual size of the carvings) in Figs. 14 and 15. Fig. 14 is from the cave of St. Michael d'Arudy, engraved on a flat piece of shoulder-bone. It shows what can only be interpreted as some kind of "halter," made apparently of twisted rope (*b, c, d*), disposed about the animal's head, whilst a broad, flat piece ornamented with angular marks is attached at the regions marked "*a*." This and other drawings similar to Fig. 13, C (of which there are many), go far to prove that these early men had mastered the horse and put a kind of bridle on his head. Fig. 15 is a solid all-round carving in reindeer's antler from the cave of Mas d'Azil, Ariège (France). The original is of this size, and is supposed to be one of the oldest and yet is the most artistic yet discovered, and worthy to compare with the horses of the Parthenon.

In Fig. 20 we have a wonderful outline of a bear engraved on a piece of stone, from the cave of Teyjat, in the Dordogne; Fig. 22, the head of a wolf on the wall of the cave of Combarelles, Dordogne; Fig. 23, lion (maneless), engraved on the wall of the same cave; Fig. 21, small bear, engraved on a pebble; Fig. 24, a duck engraved on a piece of reindeer's antler (Gourdan, Haute Garonne); Fig. 17, the square-mouthed, two-horned rhinoceros, drawn in red (ochre) outline on the wall of the cavern of the Font de Gaume. This drawing is 2-1/2 ft. long. In successful characterization the bear (Fig. 20), the wolf (Fig. 22), and the feline (Fig. 23) far surpass any of the attempts at animal drawing made by modern savages, such as the Bushmen of South Africa, Californian Indians, and Australian black fellows.

FIG. 25.—Female figure carved in oolitic limestone from Willendorf near Krems, Lower Austria (1908). Half the size (linear) of the original.

Fig. 27 is an outline sketch of a rock-carved statue, 18 in. high, proved by the kind of flint implements found with it to be of Aurignacian age. It was discovered on a rubble-covered face of a rock-cliff at Laussel, in the Dordogne, by M. Lalanne. The woman holds a bovine horn in her right hand. The face is obliterated by "weathering." Four other human statues were found in the same place, one a male, much broken, but obviously standing in the position taken by (Fig. 28) a man throwing a spear or drawing a bow. [3] Near these were found a frieze of life-sized horses carved in high relief on the rock. These are the only statues of any size, executed by the Reindeer men, yet discovered.

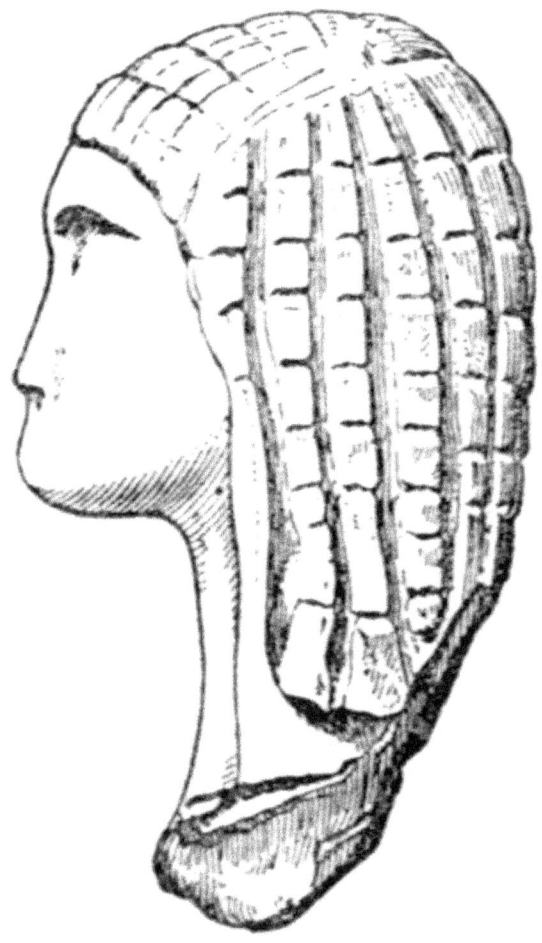

FIG. 26.—Drawing (of the actual size of the original) of an ivory carving (fully rounded) of a female head. The specimen was found in the cavern of Brassempouy, in the Landes. It is of the earliest Reindeer period, and the arrangement of the hair or cap is remarkable.

The representations of men are rare among these earliest works of art, and less successfully carried out than those of animals. But several small statuettes of women in bone, ivory, and stone of the early Aurignacian horizon are known. They suggest, by their form of body, affinity with the Bushmen race of to-day (Fig. 25). The all-round carving of a female head (Fig. 26) also suggests Ethiopian affinities in the dressing of the hair. Some regard this hair-like head-dress as a cap. Here and there badly executed outline engravings of men, some apparently wearing masks, have been discovered.

The fact that the "Reindeer men" were skilful in devising decorative design—not representing actual natural objects—is shown by the carving drawn in Fig. 29 and in many others like it.

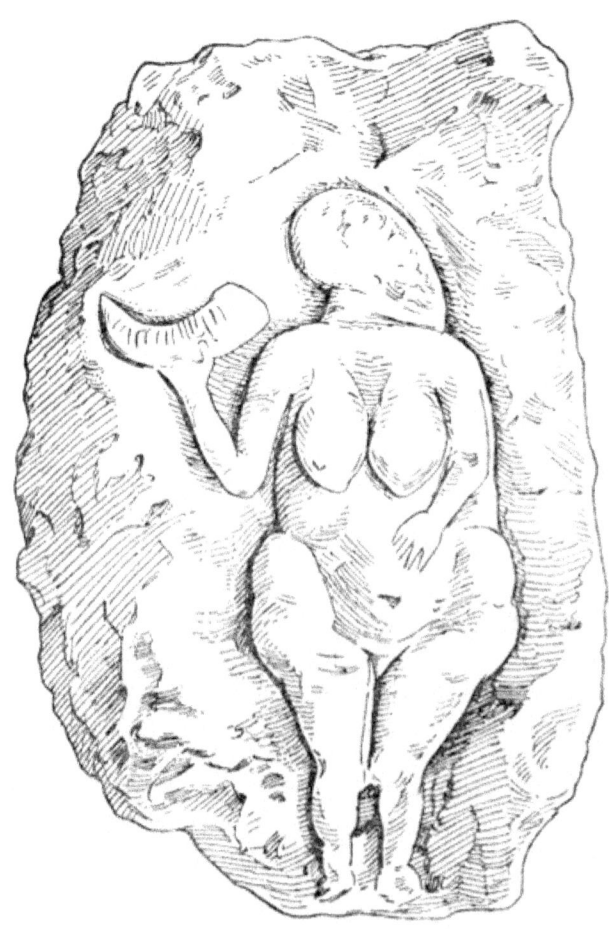

FIG. 27.—Seated figure of a woman holding a bovine horn in the right hand; high relief carved on a limestone rock; about 18 inches high. Discovered at Laussel (Dordogne) in a rock-shelter in 1911, by M. Lalanne.

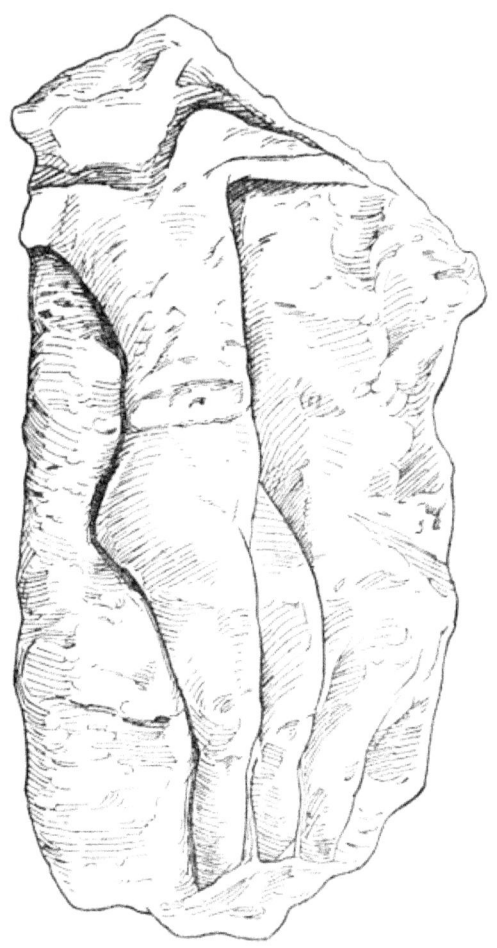

FIG. 28.—Male figure represented in the act of drawing a bow or throwing a spear. Carved on limestone rock; about 16 inches high. Discovered by M. Lalanne with that drawn in Fig. 27.

The later horizons of the Reindeer period or Upper Pleistocene yield some beautiful outline engravings of red deer and reindeer (Fig. 16) on antler-bone, as well as of other animals. One celebrated carving I have described in the first chapter of this book. It is now regarded as probable that whilst the art of the Aurignacians persisted and developed in the South of France and North-West of Spain until and during the time of the Magdalenian horizon, yet a distinct race, with a different style of art, spread through South-East Spain and also from Italy into that region, and affected injuriously the "naturalistic" Aurignacian art, and superseded it in Azilian and Neolithic times. We find late drawings (Azilian age?) in some of the east Spanish caves of a very much simplified character, small human figures armed with bow and arrow, and others reduced to geometric or mere symbolic lines derived from human and animal form (see Fig. 52, p. 206). The latest studies of Breuil on this subject tend to throw light by aid of these simplified inartistic and symbolic drawings on the migrations of very early races in the south and south-east of Europe, and to connect them perhaps with North African contemporary races. The subject is as difficult as it is fascinating. Those who wish to get to the original sources of information should consult the last ten years' issues of the invaluable French periodical called "L'Anthropologie," edited by Professor Marcelin Boule.

FIG. 29.—A piece of mammoth ivory carved with spirals and scrolls from the cave of Arudy (Hautes Pyrénées). Same size as the object.

FOOTNOTE:

[3] M. Reinach relates ("Repertoire de l'Art Quatermaire") that two of these statues were in 1912 deliberately stolen by the German Verworn professor of Physiology in Bonn, who repaid the hospitality of M. Lalanne by bribing his workman and secretly carrying off these valuable specimens to Germany, where (it is stated) they were sold to the museum of Berlin for a large sum.

CHAPTER IV

VESUVIUS IN ERUPTION

AT intervals of ten to twenty years the best-known volcano in the world—Vesuvius, on the Bay of Naples—has in the last two centuries burst into eruption, and the probability of the recurrence of this violent state of activity, at no distant date, render some account of my own acquaintance with that great and wonderful thing seasonable. We inhabitants of the West of Europe have little personal experience of earthquakes, and still less of volcanoes, for there is not in the British Islands even an "extinct" volcanic cone to remind us of the terrible forces held down beneath our feet by the crust of the earth. In regions as near as the Auvergne of Central France and the Eiffel, close to the junction of the Moselle with the Rhine, there are complete volcanic craters whose fiery origin is recognized even by the local peasantry. They are, however, regarded by these optimist folk as the products of ancient fires long since burnt out. The natives have as little apprehension of a renewed activity of their volcanoes as we have of the outburst of molten lava and devastating clouds of ashes and poisonous vapour from the top of Primrose Hill. Nevertheless, the hot springs and gas issuing from fissures in the Auvergne show that the subterranean fires are not yet closed down, and may at any day burst again into violent activity.

Such also was the happy indifference with which from time immemorial the Greek colonists and other earlier and later inhabitants of the rich and beautiful shores of the Neapolitan bay before the fateful year A.D. 79, had regarded the low crater-topped mountain called Vesuvius or Vesbius, as well as the great circular forest-grown or lake-holding cups near Cumæ and the Cape Misenum, at the northern end of the bay—known to-day as the Solfatara, Astroni, Monti Grillo, Barbaro, and Cigliano—and the lakes Lucrino, Averno, and Agnano. These together with the Monte Nuovo—which suddenly rose from the sea near Baiæ in 1538 and as suddenly disappeared—constitute "the Phlegræan fields." Vesuvius was loftier than any one of the Phlegræan craters, and the gentle slope by which it rose from the sea level to a height of nearly 3700 ft. had, as now, a circumference of ten miles. It did not terminate in a "cone," as in later ages, but in a depressed, circular, forest-covered area measuring a mile across, which was the ancient crater. A drawing showing the shape of the mountain at this period is the work of the late Prof. Phillips of Oxford (Fig. 30). The soil formed around and upon the ancient lava-streams of Vesuvius appears to have been always especially fertile, so that flourishing towns and villages occupied its slopes, and the ports of Herculaneum, Pompeii, and Stabiæ were the seats of a busy and long-established population. The existence of active volcanoes at no great distance from Vesuvius was, however, well known to the ancient Greeks and Romans. The great Sicilian mountain, Etna—more than 10,000 ft. in height, rising from a base of ninety miles in circumference—and the Lipari Islands, such as Stromboli and Volcano, were for many centuries in intermittent activity before the first recorded eruption of Vesuvius—that of A.D. 79—and great eruptions are recorded as having occurred in the mountain mass of the island of Ischia, close to the Bay of Naples, in the fifth, third, and first centuries B.C.

FIG. 30.—Vesuvius as it appeared before the eruption of August 24, A.D. 79. From a sketch by Prof. Phillips, F.R.S.

Nevertheless, the outburst of Vesuvius in A.D. 79 and its re-entrance into a state of activity came upon the unfortunate population around it as an absolutely unexpected thing. At least a thousand years—probably several thousand years—had passed since Vesuvius had become "extinct." All tradition of its prehistoric activity had disappeared, though the learned Greek traveller Strabo had pointed out the indications it presented of having been once a seat of consuming fire. From A.D. 63 there were during sixteen years frequent earthquakes in its neighbourhood, which, as we know by records and inscriptions, caused serious damage to the towns around it, and then suddenly, on the night of Aug. 24, A.D. 79, vast explosions burst from its summit. A huge black cloud of fine dust and cinders, lasting for three days, spread from it for twenty miles around, streams of boiling mud poured down its sides, and in a few hours covered the city of Herculaneum, whilst a dense shower of hot volcanic dust completely buried the gay little seaside resort known as

Pompeii. Many thousand persons perished, choked by the vapours or overwhelmed by the hot cinders or engulfed in the boiling mud.

The great naturalist Pliny was in command of the fleet at Cape Misenum, and went by ship across the bay to render assistance to the inhabitants of the towns at the foot of Vesuvius. Pliny's nephew wrote two letters to the historian Tacitus, giving an account of these events and of the remarkable courage and coolness of his uncle, who, after sleeping the night at Stabiæ, was suffocated by the sulphurous vapours as he advanced into the open country near the volcano. The friends who were with him left him to his fate and made their escape. The younger Pliny had prudently remained, out of danger, with his mother at Misenum.

The alternating periods of activity and of rest exhibited by volcanoes seem to us capricious, and even at the present day are not sufficiently well understood to enable us to discern any order or regularity in their succession. Vesuvius is a thousand centuries old, and we have only known it for thirty. We cannot expect to get the time-table of its activities on so brief an acquaintance. Strangely enough, Vesuvius, having, after immemorial silence, spasmodically burst into eruption and spread devastation around it, resumed its slumber for many years. There is no mention of its activity for 130 years after A.D. 79. Then it growled and sent forth steam and cinder-dust to an extent sufficient to attract attention again; its efforts were thereafter recorded once or so in a century, though little, if any, harm was done by it. In A.D. 1139 there was a great throwing-up of dust and stones, with steam, which reflected the light of molten lava within the crater, and looked like flames. And then for close on 500 years there was little, if any, sign of activity. The "eruptions" between that of A.D. 79 and that of A.D. 1139 had been ejections of steam and cinders, unaccompanied by any flow or stream of lava. Then suddenly the whole business shut up for 500 years, and after that—also quite suddenly—in 1631, a really big eruption took place, exceeding in volume the catastrophe of Pliny's date. Not only were columns of dust and vapour ejected to a height of many miles, but several streams of white-hot lava overflowed the edge of the crater and reached the seacoast, destroying towns and villages on the way. Some of these lava-streams were five miles broad, and can be studied at the present day. As many as 18,000 persons were killed.

There were three more eruptions in the seventeenth century, and from that date there set in a period of far more frequent outbursts, which have continued to our own times. In the eighteenth century there were twenty-three distinct eruptions, lasting each from a few hours to two or three days, and of varying degrees of violence—a vast steam-jet forcing up cinders and stones from the crater into the air, usually accompanied by the outflow of lava, from cracks in sides of the crater, in greater or less quantity. In the nineteenth century there were twenty-five distinct eruptions, the most formidable of which were those of 1822, 1834, and 1872. All of the eruptions of Vesuvius in the last 280 years have been carefully described, and most of them recorded in coloured pictures (a favourite industry of the Neapolitans), showing the appearance of the active volcano both by day and night and its change of shape in successive years. Sir William Hamilton, the British Ambassador at the Court of Naples at the end of the eighteenth century (of whose great folio volumes I am the fortunate possessor), largely occupied himself in the study and description of Vesuvius, and published illustrations of the kind mentioned above, showing the appearance of the mountain at various epochs. Since his day there has been no lack of descriptions of every succeeding eruption, and now we have the records of photography.

The crater or basin formed by a volcano starts with the opening of a fissure in the earth's surface communicating by a pipe-like passage with very deeply-seated molten matter and steam. Whether the molten matter thus naturally "tapped" is only a local, though vast, accumulation, or is universally distributed at a given depth below the earth's crust, and at how many miles from the surface, is not known. It seems to be certain that the great pressure of the crust of the earth (from five to twenty-five miles thick) must prevent the heated matter below it from becoming either liquid or gaseous, whether the heat of that mass be due to the cracking of the earth's crust and the friction of the moving surfaces as the crust cools and shrinks, or is to be accounted for by the original high temperature of the entire mass of the terrestrial globe. It is only when the gigantic pressure is relieved by the cracking or fissuring of the closed case called "the crust of the earth" that the enclosed deep-lying matter of immensely high temperature liquefies, or even vaporizes, and rushes into the up-leading fissure. Steam and gas thus "set free" drive everything before them, carrying solid masses along with them, tearing, rending, shaking "the foundations of the hills," and issuing in terrific jets from the earth's surface, as through a safety valve, into the astonished world above. Often in a few hours they choke their own path by the destruction they produce and the falling in of the walls of their briefly-opened channels. Then there is a lull of hours, days, or even centuries, and after that again, a movement of the crust, a "giving" of the blockage of the deep, vertical pipe, and a renewed rush and jet of expanding gas and liquefying rock.

The general scheme of this process and its relations to the structure and properties of the outer crust and inner mass of the globe is still a matter of discussion, theory and verification; but whatever conclusions geologists may reach on these matters, the main fact of importance is that steam and gases issue from these fissures with enormous velocity and pressure, and that "a vent" of this kind, once established, continues, as a rule, to serve intermittently for centuries, and, indeed, for vast periods to which we can assign no definite limits. The solid matter ejected becomes piled up around the vent as a mound, its outline taking the graceful catenary curves of rest and adjustment to which are due the great beauty of volcanic cones. The apex of the cone is blown away at intervals by the violent blasts issuing from the vent, and thus we have formed the "crater," varying in the area enclosed by its margin and in the depth and appearance of the cup so produced. At a rate depending on the amount of solid matter ejected by the crater, the mound will grow in the course of time to be a mountain, and often secondary craters or temporary openings, connected at some depth with the main passage leading to the central vent, will form on the sides of the mound or mountain. Sometimes the old crater will cease to grow in consequence of the blocking of its central vent and the formation of one or more subsidiary vents, the activity of which may blast away or smother the cup-like edge of the first crater.

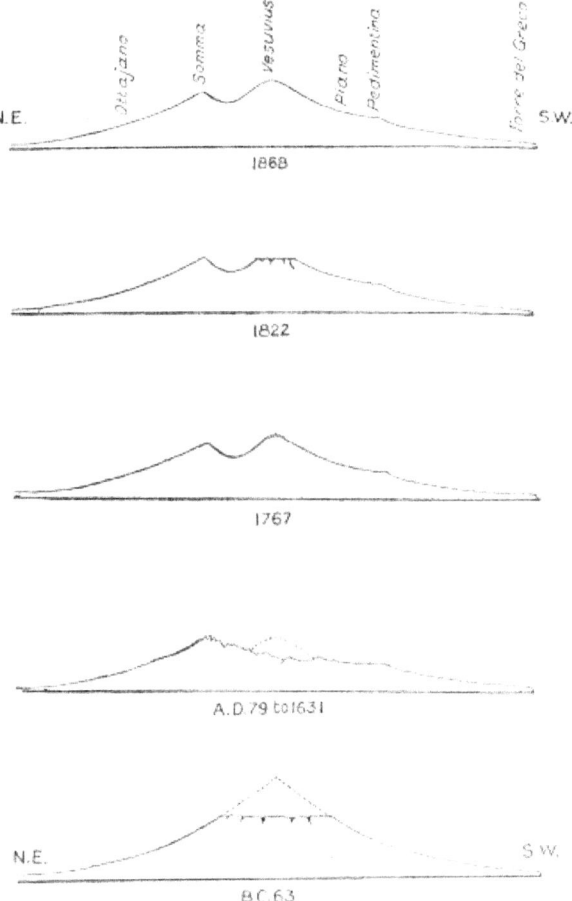

FIG. 31.—Five successive stages in the change of form of Vesuvius (after Phillips' "Vesuvius," Oxford, 1869). In the oldest (lowest figure) we see the mountain with its still earlier outline completed by the cone drawn in dotted line. Within the period of historic record that cone had not been seen. The mountain had, so far as men knew, always been truncated as shown here and in Fig. 30. The next figure above shows the further lowering of the mountain by the first eruption on record—that which destroyed Pompeii in A.D. 79. The commencing formation of a new ash-cone is indicated by a dotted line. In the three upper figures we trace the gradual growth of the new cone from 1631 to 1868. In 1872 the top of the new ash-cone was blown away, and the mountain reverted to the shape of 1822. Now (1920) the cone has accumulated once more and is higher than it was in 1868.

Such a history has been that of Vesuvius shown in outline in Fig. 31. In geologic ages—perhaps some thousands of centuries ago—Vesuvius was probably a perfect cone (its outline is shown at the bottom of p. 62) some 7000 ft. high, rising by a characteristically accelerated upgrowth from a circle of ten miles or more in diameter to its delicate central peak, hollowed out at the summit by a small crater a couple of hundred yards across. Its eruptions at that time were neither excessive nor violent. Then came a period of greatly increased energy—the steam-jet blew with such violence that it shattered and dispersed the cone, lowering the mountain to 3700 ft. in height, truncating it and leaving a proportionately widened crater of a mile and a half in diameter. And then the mountain reposed for long centuries. We do not know how long this period of extinction was, for we do not know when it began, but we know that this was the state of the mountain when in A.D. 79 it once more burst into life. In recent years—that is, since the seventeenth century A.D., a curious change took place in the mountain: the vent or orifice of the conducting channel by which eruptive matters were brought to the surface ceased to be in the centre of the wide broken-down crater of Pliny's time, and a vent was formed a few hundred yards to the south of the centre of the old crater, nearer to the south side of the old crater's wall. From this ashes or cinders issued, and were piled up to form a new cone, which soon added 600 ft. to the height of the mountain and covered in the southern half of the old crater's lip, whilst leaving the northern half or semicircle free. This latter uncovered part was called by the Italians "Monte Somma," and the new cone low down in the southern side of which the rest of the old crater-lip could be traced, was henceforth spoken of as "the ash-cone" and sometimes misleadingly as "the true" Vesuvius. Clearly it was not "the true Vesuvius" since it was a new growth. The original old Vesuvius was crowned by a crater formed by the cliffs of Monte Somma and their continuation round to the south side, now more or less completely concealed by the new ash-cone.

In the course of various eruptions during the last two centuries the new ash-cone thus formed was blown away more or less completely, and gradually grew up again. During the nineteenth century it was a permanent feature of the mountain, though a good deal cut down in 1822, and later grew so high as to give a total elevation from the sea-level of 4300 ft. The crater at the top of the ash-cone has varied during the past century in width and depth, according to its building up or blowing away by the central steam jet. In 1822 it is reported to have been funnel-like and 2000 ft. deep, tapering downwards to the narrow fissures which are the actual vent. At other times it has been largely filled by débris, and only 200 ft. deep. Molten lava has often issued from fissures in the sides of the ash-cone, and even lower down on the sides of the mountain, and a very small secondary crater has sometimes appeared on the side of the ash-cone 100 ft. or 200 ft. from the terminal crater which "finishes off" the cone.

Such was the condition of the mountain when I first saw it in the autumn of 1871. Six months later I witnessed the most violent eruption of the nineteenth century. Vesuvius kept up a continuous roar like that of a railway engine letting off steam when at rest in a covered station only a thousandfold bigger. Its vibrations shook with a deep musical note, for twenty-four hours, the house nine miles distant in Naples in which I was staying. My windows commanded a view of the mountain, and when the noise ceased and the huge steam-cloud cleared away, I saw a different Vesuvius, the higher part of the ash-cone was gone, and a huge gap in it had been formed by the blowing away of its northern side.

In October 1871, when I joined my friend Anton Dohrn at Naples in order to study the marine creatures of the beautiful bay, Vesuvius was in the proud possession of a splendid cone, completing its graceful outline. A little steam-cloud hung about one side of the cone during the day, and as night came on Vesuvius used, as we said, to "light his cigar." In fact, a very small quantity of molten lava was at that time flowing from the side of the ash-cone, about 100 ft. from its summit, and this gave a most picturesque effect as we watched it from our balcony high up on Pausilippo, when the sun set. It was a friendly sort of beacon, far away on the commanding mountain's top, which was answered by the lighting up of a thousand lamps along the coast, and by innumerable flaming faggots in the fishermen's boats moving across the bay, drawing to their light strange fishes, to be impaled by the long tridents of the skilful spearmen. That little beacon light on Vesuvius increased in volume in the course of three weeks, and was supplemented by other flaming streams and by showers of red-hot stones from the crater. This small "eruption" was the precursor by six months of the great eruption of the end of April 1872, and I spent a night on Vesuvius during its progress, and looked into the crater from which the glowing masses of rock were being belched forth.

Not long before I went, in 1871, to Naples I had spent some weeks in visiting the extinct volcanoes of the Auvergne and of the Eiffel, and I was eager to examine the still living Vesuvius. In the first week of October I made an excursion to the crater of Vesuvius in company with the son of a Russian admiral, whose name, "Popoff," was under the circumstances unpleasantly suggestive. We examined some black slaglike masses of old lava-streams, and struggled up the loose sandy ash-cone (there was no "funicular" in those days), and prodded with our sticks the few yards of molten lava which emerged from the side of the cone about 100 ft. from the summit. On Nov. 1 my friend Anton Dohrn (who was then negotiating with the Naples Municipality for a site in the Villa Nazionale on which to erect the great Zoological Station and Aquarium, now so well known) was with me and some Neapolitan acquaintances looking at Vesuvius across the bay from Pausilippo, where we had established ourselves, when we noticed that a long line of steam was rising from the lower part of the ash-cone and that puffs of steam were issuing at intervals from the crater. "Dio mio! Dio di Dio!" cried the Neapolitans in terror, and expressed their intention of leaving Naples without an hour's delay. As night fell a new glowing line of fire appeared far down near the base of the ash-cone, whilst what looked in the distance like sparks from a furnace, but were really red-hot stones—each as big as a Gladstone bag—were thrown every two or three minutes from the crater.

We hired a carriage, drove to Resina (built above buried Herculaneum), and walked up towards the Observatory in order to spend the night on the burning mountain. We found that two white-hot streams, each about twenty yards broad at the free end, were issuing from the base of the cone. The glowing stones thrown up by the crater were now separately visible; a loud roar accompanied each spasmodic ejection. The night was very clear, and a white firmly-cut cloud, due to the steam ejected by the crater, hung above it. At intervals we heard a milder detonation—that of thunder which accompanied the lightning which played in the cloud, giving it a greenish illumination by contrast with the red flame colour reflected on to it by red-hot material within the crater. The flames attributed to volcanoes are generally of this nature, but actual flames do sometimes occur in volcanic eruptions by the ignition of combustible gases. The puffs of steam from the crater were separated by intervals of about three minutes. When an eruption becomes violent they succeed one another at the rate of many in a second, and the force of the steam jet is gigantic, driving a column of transparent super-heated steam with such vigour that as it cools into the condition of "cloud" an appearance like that of a gigantic pine-tree seven miles high (in the case of Vesuvius) is produced.

We made our way to the advancing end of one of the lava-streams (like the "snout" of a glacier), which was 20 ft. high, and moved forwards but slowly, in successive jerks. Two hundred yards farther up, where it issued from the sandy ashes, the lava was white-hot and running like water, but it was not in very great quantity and rapidly cooled on the surface and became "sticky." A cooled skin of slag was formed in this way, which arrested the advancing stream of lava. At intervals of a few minutes this cooled crust was broken into innumerable clinkers by the pressure of the stream, and there was a noise like the smashing of a gigantic store of crockery ware as the pieces or "clinkers" fell over one another down the nearly vertical "snout" of the lava-stream, whilst the red-hot molten material burst forward for a few feet, but immediately became again "crusted over" and stopped in its progress. We watched the coming together and fusion of the two streams and the overwhelming and burning up of several trees by the steadily, though slowly, advancing river of fire. Then we climbed up the ash-cone, getting nearer and nearer to the rim of the crater, from which showers of glowing stones were being shot. The deep roar of the mountain at each effort was echoed from the cliffs of the ancient mother-crater, Monte Somma, and the ground shook under our feet as does a ship at sea when struck by a wave. The night was very still in the intervals. The moon was shining, and a weird melancholy "ritornelle" sung by peasants far off in some village below us came to our ears with strange distinctness. It might have been the chorus of the imprisoned giants of Vulcan's forge as they blew the sparks with their bellows and shook the mountains with the heavy blows of their hammers.

As we ascended the upper part of the cone the red-hot stones were falling to our left, and we determined to risk a rapid climb to the edge of the crater on the right or southern side, and to look into it. We did so, and as we peered into the great steaming pit a terrific roar, accompanied by a shuddering of the whole mountain, burst from it. Hundreds of red-hot stones rose in the air to a height of 400 ft., and fell, happily in accordance with our expectation, to our left. We ran quickly down the sandy side of the cone to a safe position, about 300 ft. below the crater's lip, and having lit our pipes from one of the red-hot "bombs," rested for a while at a safe distance and waited for the sunrise. A vast horizontal layer of cloud had now formed below us, and Vesuvius and the hills around Naples appeared as islands emerging from a sea. The brilliant sunlight was reassuring after this night of strange experiences. The fields and lanes were deserted in the early morning as we descended to the sea-level. On our way we met a procession of weird figures clad in long white robes, enveloping the head closely but leaving apertures for the eyes. They were a party of the lay-brothers of the Misericordia carrying a dead man to his grave. Then we found our carriage, and drove quickly back to Naples and sleep!

In the following March I acted as guide to my friend Professor Huxley in expeditions up Vesuvius, now quiescent, and to the Solfatara. Then suddenly, in April, the great eruption of 1872 burst upon us. On the first day of the outbreak some imprudent visitors were killed by steam and gas ejected by the lava-stream. By the next day the violence of the eruption was too great for any one to venture near it. The crater sent forth no intermittent "puffs" as in the preceding November, but a continuously throbbing jet which produced a cloud five miles high, like an enormous cauliflower in shape, suspended above the mountain and making it look by comparison like a mole-hill. Showers of fine ashes, as in the days of Pompeii, fell thickly around, accumulating to the depth of an inch in a few hours even at my house in Pausilippo, nine miles distant across the bay. I was recovering at the time from an attack of typhoid fever, and lay in bed, listening to the deep humming sound and wondering at the darkness until my doctor came and told me of the eruption. I was able to get up and see from the window the great cauliflower-like cloud and the vacant place where the ash-cone was, but whence it had how been scattered into the sky. (It has been gradually re-formed by later eruptions, of which the last of any size was in 1906.) I could also see steam rising like smoke from a long line reaching six miles down the mountain into the flat country below. It was the great lava-stream which had destroyed two prosperous villages in its course.

After ten days I was able to get about, and drove over to one of these villages and along its main street, which was closely blocked at the end by what looked like a railway embankment some 40 ft. high. This was the side of the great lava-stream now cooled and hardened on the surface. It had sharply cut the houses, on each side of the street, in half without setting them on fire, so that the various rooms were exposed in section—pictures hanging on the walls, and even chairs and other furniture remaining in place on the unbroken portion of the floor. The villagers had provided ladders by which I ascended the steep side of the embankment-like mass at the end of the street, and there a wonderful sight revealed itself. One looked out on a great river seven miles long, narrow where it started from the broken-down crater, but widening to three miles where I stood, and wider still farther on as it descended. This river, with all its waves and ripples, was turned to stone, and greatly resembled a Swiss glacier in appearance. A foot below the surface it was still red-hot, and a stick pushed into a crevice caught fire. It was not safe to venture far on to the pie-crust surface. A couple of miles away the campanile of the church of a village called Massa di Somma stood out, leaning like that of Pisa, from the petrified mass, whilst the rest of the village was overwhelmed and covered in by the great stream.

The curious resemblance of the lava-stream to a glacier arose from the fact that it was almost completely covered by a white snow-like powder. This snow-like powder, which often appears on freshly-run lava, is salt—common sea salt and other mineral salts dissolved in the water ejected as steam mixed with the lava. The steam condenses, as the lava cools, into water and evaporates slowly, leaving the salt as crystals. Often these are not white, but contain iron salt, mixed with the white sodium, potassium, and ammonium chlorides, which gives them a yellow or orange colour. Salts coloured in this way have the appearance of sulphur, and are often mistaken for it. The whole of the interior of the crater of Vesuvius when I revisited it in 1875 was thus coloured yellow, and I have a water-colour sketch of the scene made by a friend who came with me for the purpose. As a matter of fact, though small quantities of the choking gas called "sulphurous acid" are among the vapours given off by Vesuvius, there is no deposit of sulphur there. Some large volcanoes (in Mexico and Japan) have made deposits of sulphur, which are dug for commercial purposes; but the sulphur of Sicily is not, and has not been, thrown out or volatilized by Etna. It occurs in rough masses and in splendid crystals in a tertiary calcareous marine deposit, and its deposition was probably due to a chemical decomposition of constituents of the sea water brought about by minute plants, known as "sulphur bacteria." Whether the neighbouring great volcano had any share in the process seems to be doubtful.

It is generally supposed that sea-water makes its way in large quantity through fissures connected with volcanic channels, and is one of the agents of the explosions caused by the subterranean molten matter. Gaseous water, hydrochloric acid, carbonic acid, hydrofluoric acid, and even pure hydrogen and oxygen and argon are among the gases ejected by volcanoes.

The molten matter forced up from the bowels of the earth and poured out by volcanoes is made up of various chemical substances, differing in different localities, and even in different eruptions of the same volcano. It consists largely of silicates of iron, lime, magnesium, aluminium, and the alkali metals, with possible admixture of nearly every other element. Some volcanoes eject "pitch" or "bitumen." When the molten matter cools, interesting crystals of various "species" (*i.e.*, of various chemical composition) usually form in the deeper part of the mass. The lavas of Vesuvius frequently contain beautiful opaque-white twelve-sided crystals of a siliceous mineral called "leucite." I have collected in the lava of Niedermendig, on the Rhine, specimens embedding bright blue transparent crystals (a mineral called Haüynite) scattered in the grey porous rock. The lava-streams, and even the "roots," of extinct volcanoes which are of great geologic age, sometimes become exposed by the change of the earth's surface, and extensive sheets of volcanic rock of various kinds are thus laid bare. Basalt is one of these rocks, and it not unfrequently presents itself as a mass of perpendicular six-sided columns, each column 10 ft. or more high, and often a foot or more in diameter. The "Giant's Causeway," in the North of Ireland, and the "Pavée des Géants," in the Ardêche of Southern France, are examples both of which I have visited. It is not easy to explain how the molten basalt has come to take this columnar structure on cooling. It has nothing to do with "crystallization," but is similar to the columnar formation shown by commercial "starch" and occasionally by "tabular flint". A theoretical explanation of its formation has been given by Prof. J. Thompson, brother of the late Lord Kelvin.

The varieties of volcanoes and their products make up a long story—too long to be told here. There are from 300 to 400 active craters in Existence to-day—mostly not isolated, but grouped along certain great lines, as, for instance, along the Andean chain, or in more irregular tracks. If we add to the list craters no longer active, but still recognizable, we must multiply it by ten. Vesuvius is the only active volcano on the mainland of Europe—Hecla, Etna, Stromboli, Volcano, and the volcanoes of the Santorin group are on islands. The biggest volcanoes are in South America, Mexico, Java, and Japan. Volcanoes and the related "earthquakes" have been most carefully studied with a view to the safety of the population in Japan. The graceful and well-beloved volcano, Fujiyama, is more than 12,000 ft. high, but, unlike others in those islands, it has been quiescent now for just 200 years. The most violent volcanic eruptions of recent times, with the largest "output" of solid matter, are those of the Soufrière of St. Vincent in 1812, of the Mont Pelée of Martinique in 1902, and of Krakatoa in 1883. A single moderate eruption of the great volcano Mauna Loa, in Hawaii, nearly 14,000 feet high, throws out a greater quantity of solid matter than Vesuvius has ejected in all the years which have elapsed since the destruction of Pompeii. Many hundred millions of tons of solid matter were ejected by Mont Pelée in 1902, when also a peculiar heavy cloud descended from the mountain, hot and acrid, charged with incandescent sand, and rolling along like a liquid rather than a vapour. It burnt up the town of St. Pierre and its inhabitants and the shipping in the harbour. In the eruption of the volcano of St. Vincent in 1812 three million tons of ashes were projected on to the Bahamas Islands, 100 miles distant, besides a larger quantity which fell elsewhere. The great explosion at Krakatoa, lasting two days, blew an island of 1400 ft. high, into the air. A good deal of it was projected as excessively fine needlelike particles of pumice with such force as to carry it up thirty miles into the upper regions of the atmosphere, where it was carried by air currents all over the world, causing the "red sunsets" of the following year. The sky over Batavia, 100 miles distant, was darkened at midday so completely that lamps had to be used—as I heard from my brother who was there at the time. The explosions were heard in Mauritius, 3000 miles away. A sea wave 50 ft. high was set going by the submarine disturbance, and reaching Java and neighbouring islands inundated the land and destroyed 36,000 persons. This wave travelled in reduced size over a vast tract of the ocean, and was observed and recorded at Cape Horn, 7800 miles distant from its seat of origin.

CHAPTER V

BLUE WATER

MOST people know and admire the splendid expanse of blue colour offered by the clear sea water on many parts of our coasts, and by that of lakes at home and abroad. I find that there is still a sort of a fixed determination not to believe that this colour is due (as it is) to the actual blue colour of pure water. Pure, transparent water is blue. Those who think they know better will point to a glass of pure water, hold it up to the light, and affirm that it is colourless. But this apparent colourlessness is due to the small breadth of water in the glass through which the light passes. It is definitely ascertained that if water as pure and as free from either dissolved or suspended matter as it is possible to make it (by distillation and the use of vessels not acted upon by water) be made to fill an opaque tube 15 ft. long, closed at each end by a transparent plate, and then a beam of light be made to traverse the length of the tube, so that the eye receives the light after it has passed through this length of 15 ft. of water, the colour of the light is a strong blue. Water is blue in virtue of its own molecular character, just as sulphate of copper is. Liquid oxygen, prepared by the use of intense cold, is also transparent blue, and the peculiar condensed form of oxygen known as "ozone" is, when liquefied, of a darker or stronger blue than oxygen.

At one time (some thirty years ago) there was still some doubt as to whether water was self-coloured blue, or whether its blue colour was due to the action on light of excessively minute solid white particles of chalk suspended in the water. Such fine suspended particles in some cases act on the light which falls on to them so as to reflect the blue rays. This occurs in certain natural objects which have a blue colour. But these can be distinguished from transparent self-coloured blue substances by the fact that whilst the light reflected from their surface is blue, the light which is made to traverse them (when they are held up to the light so that they come between one's eye and the sun's rays) is brown. This is the case with very hot smoke, and can be well seen when a cigar is smoked in the sunlight. The smoke which comes off from the lighted end of the cigar is very hot, and its particles are more minute than those of cooler smoke. The hot smoke shows a bright blue colour when the sunlight falls on it and is reflected, but when you look through the smoke-cloud at a surface reflecting the sunlight, the cloud has a reddish-brown tint. As the smoke cools its particles adhere to one another and form larger particles, and the light reflected from the cloud is no longer blue but grey, and even white. Thus the smoke which the smoker keeps for half a minute in his mouth is cooled and condensed, and reflects white light—is, in fact, a white cloud—when he puffs it out, and contrasts strongly with the blue cloud coming off from the burning tobacco at the lighted end of the cigar. The blue colour of the sky is held by many physicists (though other views have been of late advanced) to be due to the same action on the part of the very finest particles of watery vapour, which are diffused through vast regions of our atmosphere above the condensed white-looking clouds consisting of larger floating particles of water.

Vapours are given off by many liquids, and even by some solids, varying in their production according to the heat applied in different cases. They are gases, and quite transparent and invisible at the proper temperature, like the atmospheric air. Thus water is always giving off "water-vapour," which is quite invisible. When water is heated to the boiling point it is rapidly converted into transparent invisible vapour. Steam, as this vapour is called, is invisible, and we all habitually make a misleading use of the word "steam" when we apply it both to this and to the slightly cooled and condensed cloud which we can see issuing from the spout of a kettle or from a railway engine. It seems that the fault lies with the scientific writers, who have applied the word "steam" to the invisible water vapour or gas before it has condensed to form a cloud. The old English word "steam" certainly means a visible cloudy emanation, and not a transparent invisible gas. A cloud is not a vapour, but is produced by the coming together or condensation of the minute invisible particles of a vapour to form larger particles, which float and hang together, and reflect the light, and thus are visible.

By the examination of other vapours or gases than that which is gaseous water, namely, the vapours of bodies like chloroform and ether, it has been shown that "cloud" forms in a vapour not merely in consequence of the cooling of the vapour, but in consequence of the presence in the air (or in the tube in which the vapour is enclosed for observation) of very fine floating dust particles. They act as centres of attraction and condensation for the vapour particles. When there are no dust particles present clouds do not form readily in cooling vapours, or only at lower temperatures, and in larger mass. Tyndall made some beautiful experiments on this subject, obtaining clouds of great tenuity in vapours enclosed in tubes, which reflected the most vivid blue tints when illuminated by the electric arc-lamp. Later Aitken, of Edinburgh, showed that the "fog" which forms in smoke-ridden towns is due to the condensation of previously invisible watery vapour as "cloud" around the solid floating particles of carbon of the smoke. Aitken further used this property of solid floating particles, namely, that they cause the formation of fog and cloud in vapours—to test the question as to whether the excessively minute odoriferous particles which affect our noses as "smell" are distinct solid floating particles as often supposed, or are of the nature of gas and vapour. He admitted strong perfumes such as musk into tubes containing watery vapour, at such a temperature that the vapour was in a "critical" state—just ready to condense and precipitate as "cloud." If he had admitted fine solid particles such as a minute whiff of smoke, or some "dusty" air—the cloud would have formed. But the admission of the perfume had no such effect Therefore, he concluded that the odoriferous emanations used by him are not distinct particles like those of smoke or dust, but are gaseous.

The beautiful blue tint of the semi-transparent "white" of a boiled plover's egg is due to a fine-particled cloud dispersed in the clear albumen. London milk used to be "sky-blue" for a similar reason, before the recent legislation against the adulteration of food. The blue eyes of our fair-haired race and of young foxes are not due to any "pigment"—that is to say, a separable self-coloured substance—but to a fine cloud floating in a transparent medium which reflects blue rays of light as blue smoke does. The iris of the eye can and often does develop a pigment, but it is a brown one. When present in small quantity it produces a green-coloured iris, the pale yellow-brown being added to the blue cloud-caused colouring. When present in larger quantity the same pigment gives us brown and so-called "black" eyes. The blue colours in birds' feathers and insects' wings are produced without blue pigment by special effects of reflection, and where green is the colour it is often due to the addition of a small quantity of yellow pigment to what would otherwise look blue: though some caterpillars and grasshoppers have a real green pigment in their skin. Flowers, on the other hand, have true soluble blue "pigments," and green ones too, notably that called leaf-green or chlorophyll. The little green tree frog has no blue or green pigment in its skin; only a yellow pigment. Sometimes rare specimens are found in which the yellow pigment is absent altogether, and then the little frog is turquoise-blue in colour. But there is no blue pigment in the skin; only a finely-clouded translucent film overlying a dead-black deep layer of the skin, and the result is that the frog is of a wonderful pure blue. Sometimes the commoner large edible frog is found with a similar absence of yellow pigment (I found some in a garden near Geneva six years ago), and then all the parts of its skin which usually are green show as brilliant blue.

It is at first difficult to believe that such fine, smoothly-spread turquoise blue as that of the blue frog is due merely to a "reflection effect," and that there is no blue pigment present which would show as blue if light were transmitted through it, or could be separated and dissolved in some medium. Yet this is undoubtedly the case. The nearest experimental production of such a blue surface without blue pigment is obtained by first varnishing a black board, and when the varnish is nearly dry passing a sponge wetted with water over it. Some of the varnish is precipitated from its solution in the spirit (or it may be turpentine) as a fine cloud, and until the water has evaporated it looks like blue paint, as the poet Goethe found when cleaning a picture. It would be interesting to know more precisely the precautions to be taken in order to get the blue colour in this way in fullest degree.

It appears that when light is reflected from a cloud of fine colourless particles so as to give a predominant blue colour, the light so reflected is affected in that special way which physicists describe as being "polarized." It is possible by the use of certain apparatus (the polariscope) to distinguish polarized from non-polarized light, so that it should be possible to decide (or at any rate to gain evidence) whether blue water—a sheet of blue water— owes its colour to fine particles suspended in it or to the self-colour of the water. An admirable case for making this simple experiment is presented by the great tanks—some 20 ft. cube—which are used by the water companies which draw their water supply from the chalk, for the purpose of precipitating the dissolved chalk—"Clarking" the water, as it is called, after the inventor of the process—and so getting rid of its excessive "hardness." Such tanks are to be seen by the side of the railway near Caterham. The water in these tanks is of such a brilliant turquoise blue that many people suppose that copper has been added to the water to free it from microbes! Such, at any rate, was the conviction expressed by a friend in conversation with me only a few weeks ago. The water in these tanks, when seen from the railway, looks like a magnificent blue dye, and a very important point for those (not a few) who believe that the blue colour of seas and lakes is due to the reflection of the blue colour of the sky overhead is that the water in the tanks looks just as blue when the sky is overcast with cloud as when there is blue sky. The blue colour of water has, as a rule, nothing to do with the reflection of the sky, though it is the fact that a shallow film of water may at a certain angle reflect the sky to our eyes, just as a mirror may. The effect is quite unlike that due to light passing through deep water when reflected from below it. If we examine the tanks in question we find that they have been filled with water pumped from the chalk, and that then lime has been added to the water in order to combine with the carbonic acid dissolved in it and form chalk or carbonate of lime—which is insoluble in pure water and falls as an excessively fine white powder to the bottom of the tank. But the important fact is that water having carbonic acid dissolved in it can dissolve carbonate of lime or chalk to a certain amount: and this water pumped from the chalk, having carbonic acid naturally dissolved in it, has consequently also dissolved a quantity of chalk. It is this which gives the chalk-spring water the objectionable quality of "hardness." When lime is added to the chalk-spring water as pumped into the tanks, the carbonic acid in it is taken up by the lime, and the chalk previously dissolved by the carbonic-acid-holding water is, so to speak, "undissolved," and thrown down as a very fine white powder, together with the chalk newly formed by the union of the lime and the carbonic acid. These large tanks are used to allow the fine powder of chalk to settle down and leave the water clear. The brilliantly white chalk sediment accumulates not only on the floor of the tank, but on its sides. Any light which falls on the tank is refracted and reflected from side to floor and from floor to side, and eventually emerges from the tank, a great deal of it having traversed the 20 ft. breadth and depth many times. Most of its red, yellow, and green rays are quenched by the many feet of blue water through which it has passed, and it issues as predominantly blue. This is largely due to the fine reflecting surface furnished by the "white-washed" or chalk-coated floor and sides and the great purity of the white reflecting material—no yellow or brown matter being present to give a greenish tinge to the result I remember being taken to see "Clark's process" in use, and the splendid blue colour of the water in the "softening" tanks at Plumstead, when the process was first used by the North Kent Water Company, sixty-four years ago.

It is, I think, still a possible question as to whether the fine floating particles of precipitating chalk act in any way as a "cloud"—in short, as the blue clouds of smoke, egg-white, milk, and varnish. There is no evidence that they do, but no one, so far as I know, has ever taken the trouble to settle the question. It could be done by examining the blue light from the tanks with a polariscope, and also by sinking a black tarpaulin into the tank to cover the white floor and hanging others at the sides. Then if the blue colour were due to light reflected from the white floor and sides traversing repeatedly the clear self-coloured blue water, the blue colour should no longer be visible, for the reflecting surfaces would be covered by the black tarpaulin and little light sent up through the water. But if it were due to a cloud of greatest delicacy in the water—like fine smoke reflecting the blue light rather than the other rays—then the colour should be as intense or more intense when the black background is introduced. I am surprised that some inquirer, younger and more active than I am, does not put the matter to the test of experiment.

On the whole, practically all the facts which we know about "blue water" are consistent with the blue self-colour of water, and not with that of a "blue cloud" in the water. Now that we have porcelain baths of the purest white and of large size, one may often see the strong blue colour of water of great purity in the bath, especially where waves or ripples send to our eyes those rays of light which have taken a more or less horizontal course from side to side of the bath, and have thus been through a large thickness of the pale-coloured fluid. Great masses of clear ice, such as one may study in glaciers, are blue; the "crevasses" which transmit light which has passed through a considerable thickness of ice (as, for instance, in an ice cave), are deep blue; there is no question of a reflection from suspended particles. The green colour which some glaciers show at a little distance is due to the yellow rust—iron oxide—blown on to the surface of the ice and dissolved. Many glaciers or parts of glaciers are quite free from it, and of a splendid indigo blue in their deeper fissures. So, too, as to the sea and lakes. The Blue Grotto or Cavern of the island of Capri, near Naples, is a case in point. All the light which enters it comes through the sea-water and is blue. I was taken to it in a boat rowed by two men. As the boat enters the low mouth of the cavern you have to bend down to avoid knocking your head against the rock. Then you find yourself floating in a vast and lofty chamber the white rocky floor of which is some twenty feet below the surface of the clear water. No light enters the cavern by the low part of the entrance above water. Below the surface it widens and the strong Southern sun shines through the clear water and its light is reflected up into the cave from the bottom. It is blue, and everything in the cave above as well as below the water is suffused with a blue glow—a truly wonderful and fascinating spectacle. In order to get the best effect you must choose an hour when the sun is in a favourable position. Where there is a white bottom at a depth of fifty or a hundred feet, the sea has a fine ultramarine colour, so long as it is clear. It is often made green by yellow-coloured impurities, either fine iron-stained sediment or by minute living things in the water. The colour of the water of either sea or lakes, when it is clear and overlying great depths (200 fathoms and more), tends to be dark indigo owing to the deficiency of reflected light. But there are enough white particles as a rule to send some of the light, which penetrates the water, upwards again. Even the great ocean has a dark purplish-blue colour, but never the bright blue of clear water in shallow seas with light-coloured or white bottom.

One of the most beautiful exhibitions of the colour of clear water in various thicknesses which I know, is at the entrance of the Rhone into the Lake of Geneva. The thick pale-coloured brownish-white sediment of the river shoots out for a quarter of a mile or more into the dark blue waters of the deep lake, and on a bright sunny day as it subsides reflects the light upwards from different depths through the clear water. Where it has sunk but little the colour is green, owing to the influence of the yellow mud. Farther on it is ultra-marine blue, and then, where it has sunk deeper, we get full indigo tints. The movement of the water and its churning up by the steamers' paddles add to the variety of effects, since the foam of air-bubbles submerged throws up the light through the water. It is not possible to doubt as one watches the admixture of the river and the lake, and the eddies and hanging walls of sediment, that one is floating over a vast depth of magnificent blue self-coloured fluid which is traversed by the sunlight in ways and degrees varying according to its depth and the volume of the pale mud of the in-rushing Rhone and the abundance of fine air-bubbles "churned" into the water by the paddle-wheels of the steamer.

CHAPTER VI

THE BIGGEST BEAST

THERE is a prevalent notion, encouraged by the fanciful exaggerations of newspaper gossips, that the animals of past ages, whose bones are from time to time dug out of rocks and sand quarries, were many of them much bigger than any at present existing, and that we are living in an age of degeneracy. It is true that the mammoth and the mastodon were enormous creatures, but they were *not* bigger than their living representatives, the great elephants of Africa and India. The African elephant often stands 11 ft. high at the shoulder, and occasionally attains 12 ft.

Some eighty years ago Dr. Gideon Mantell became celebrated by his discovery of the bones of huge reptiles—far bigger than any existing crocodile or lizard—nearly as big as elephants, in the Wealden rocks of Tilgate Forest in Sussex. He and Sir Richard Owen distinguished several kinds—the Iguanodon, the Megalosaurus, the Hylæosaurus, and others. Models of these creatures as they appeared when clothed in flesh and hide were carefully made, and placed picturesquely among the ponds and islands of the gardens of the Crystal Palace at Sydenham when it was first opened to an enchanted public in the fifties. As a small boy I, at that time, fell under their spell.

The passing years have brought to us more complete knowledge of these strange beasts—now classed as the "Dinosauria"—and new kinds and complete skeletons of those already known have been discovered in the United States and in Belgium. The leg bones and vertebræ of one of the biggest were found near Oxford, and are in the Oxford Museum; it received the name Cetiosaurus. Only a few years ago a very complete skeleton of a creature closely allied to Cetiosaurus was with great labour and skill dug out of the Jurassic rocks of Wyoming, U.S.A., by Dr. Holland, at the charges of Mr. Andrew Carnegie. It was known as Diplodocus (referring to certain bones in its tail), and a wonderful cast of the completely reconstructed skeleton was presented to the Natural History Museum in London, when I was Director, by Mr. Carnegie. The skeleton is 84 ft. long; but we must not be mis-led as to the animal's actual bulk by this measurement, for the tail is 46 ft. long and whip-like, whilst the neck is 23 ft. long and carries a small head not bigger than that of a cart-horse. The jaws were provided with small peg-like teeth, showing that the beast fed on soft vegetable matter. The body, apart from neck and tail, was really only a little bigger than that of a large elephant, and the limb-bones longer in the proportion of about six to five. Another reptile very similar to these and also found in the mesozoic rocks of the U.S. America is Brontosaurus.

The fact is that, if we wish to make an intelligent comparison of the sizes of different animals, we have carefully to ascertain not merely the length measurements, but the *proportions* of the various parts, and the actual bulk and probable weight of the beasts under consideration. Also (and this is a very important and decisive matter) we must know whether the beasts were terrestrial in habit, walking with their bodies raised high on their legs, or whether they were aquatic and swam in the lakes or seas, their bodies buoyed up and supported by the water. By far the biggest animals of which we have any knowledge are the various kinds of whales still flourishing in the sea. A mechanical limit is set to the size of land-walking animals, and that limit has been reached by the elephant "Flesh and blood," and we may add "bone," cannot carry on dry land a greater bulk than his. He is always in danger of sinking by his own weight into soft earth and bog. His legs have to be much thicker in proportion than those of smaller animals—made of the same material—or they would bend and snap. His feet have to be padded with huge discs of fat and fibre to ease the local pressure, and his legs are kept straight not bent at the joints, when he stands (a fact to which Shakespeare makes Ulysses refer), so that the vast weight of his body shall be supported by the stiff column formed by the upper and lower half of the limb-bones kept upright in one straight line. A well-grown elephant weighs five tons. Compare his weight and shape with that of a big whale-bone whale! No extinct animal known approaches the existing whale in bulk and weight. He is 80 to 90 ft. long, and has no neck nor any length of tail. His outline is egg-like, narrower at the hinder end. He weighs 200 tons—forty times as much as a big elephant—and is perfectly supported without any strain on his structure by the water in which he floats. There is no such limit to his possible size as there is in the case of land-walking animals. But it seems probable that he too is limited in size by mechanical conditions of another kind. Probably he cannot exceed some 90 ft. in length and 200 tons of bulk on account of the relatively great increase of proportionate size and power in the heart required in order to propel the blood through such a vast mass of living tissue and keep him "going" as a warm-blooded mammal. The original pattern—the small dog-like ancestor of the whale—cannot be indefinitely expanded as an efficient working machine, though its limit of growth is not determined by the same mechanical causes as those which limit the bulk of the terrestrial quadruped.

These considerations make it clear that we should compare as to "bigness" terrestrial animals with other terrestrial animals, and aquatic animals with aquatic ones. It seems probable that Diplodocus was an aquatic reptile, and never raised himself on to his four legs on dry land as the Carnegie skeleton at the Natural History Museum is doing. His legs and feet are quite unfitted to support his weight on a land surface; on land he would have rested on his belly, as a crocodile does, with much bent legs on each side. But submerged in 20 ft. depth of water, he could have trotted along, half-floating, with his feet touching the bottom and his head raised on its long neck to the surface, slowly sucking the floating vegetation into his moderate-sized mouth. (See drawing on p. 91.)

Diplodocus and Cetiosaurus have huge thigh-bones and upper-arm bones—respectively 5 ft. 9 in. and 3 ft. 2 in. in length—until lately the biggest known *limb*-bones, although the lower jaw-bone of a Right Whale grows to be 18 ft. in length. But a thigh-bone (femur) of a reptile similar to Diplodocus has been found in Wyoming, 6 ft. 2 in. in length. This reptile was named Atlantosaurus, and a cast of the huge bone—the biggest known when it was placed there—stands in our museum gallery. However, its glory has departed, for we now know "than this biggest bone, a bigger still." The bones of several individuals of a huge reptile similar to Diplodocus, but actually twice as big in linear dimensions, were found by Dr. Fraas at Tendagoroo, fifty miles from the coast in German East Africa, and brought safely to Berlin in 1912, though they have not yet been mounted as a complete specimen. They were lying in a sandy deposit of the same geologic age as our Sussex Wealden. A special expedition of 500 negroes was sent—not by the Government, but by the Berlin "Society of the Friends of Natural History" (we need such a society in England), at a cost of £10,000, to fetch the bones. They were of many individuals, and had to be skilfully dug out and packed. Dr. Fraas calls this biggest of all quadrupeds "Gigantosaurus." A cast of the humerus, or upper-arm bone, is now exhibited in the Natural History

Museum. It is over 7 ft. in length. The femur, or thigh-bone, was still bigger—it was over 10 ft. in length. Alas for the glory of Atlantosaurus! This enormous creature was, of course, like Diplodocus, aquatic. Its bulk was much less than that of a big whale, but extinct aquatic reptiles may yet be found of greater size. Ichthyosaurus, the extinct whale-like reptile, does not exceed 30 ft. in length. Our engraving (Fig. 32) shows the relative size of the humerus of man, the elephant, [4] and the Gigantosaurus. How puny is that human arm-bone! And yet...!

When stretched on the shore, resting on the belly, the body of the great lizard of Tendagoroo bulked like a breakwater 12 ft. high, and his tail like a huge serpent extended 80 ft. beyond it; whilst his head and neck reached 40 ft. along the mud in front.

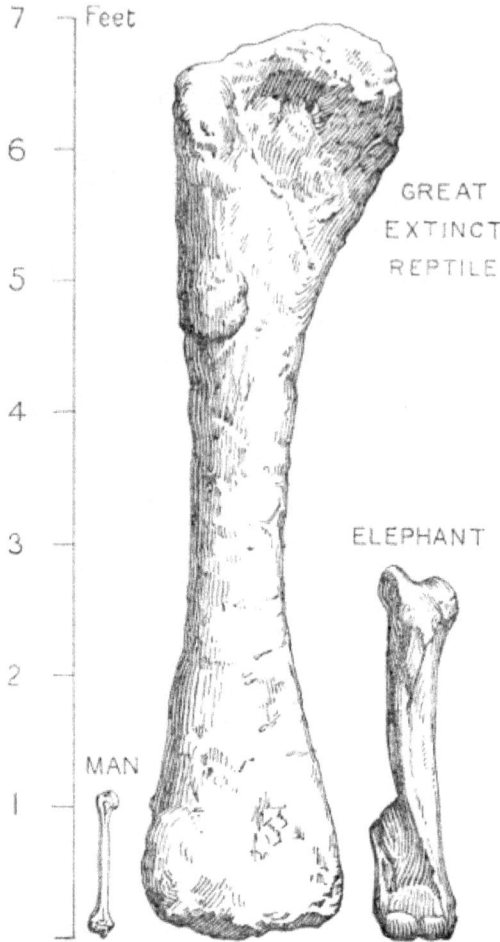

FIG. 32.—The upper-arm bone or humerus of the great reptile (Gigantosaurus) of Tendagoroo—compared with that of man and of an Indian elephant.

An important limitation to great size in an animal is, it must be remembered, often imposed by the nature of the animal's food. Ten individuals each weighing a hundredweight will more easily pick up and swallow the amount of food required to nourish ten hundredweight of the species than will one individual responsible for the whole bulk, provided that the food is scattered and not ready to the mouth in unlimited quantity. A creature which has unlimited forest or grass or seaweed as its food will be at no disadvantage owing to its size. But a carnivor or a fish-eater or one depending on special fruits and roots not offered to him by nature in mass has to search for, and sometimes to hunt, or at any rate to compete with others, for the scattered and elusive "bits" of food. So it is that we find that the fruit-eating apes are not very big, and that terrestrial carnivors are small, though powerful and swift, as compared with cattle, deer, and vegetarian beasts. Ten carnivors weighing each ten stone will with their ten mouths "pick up" more prey than one carnivor weighing a hundred stone and having only one mouth. Even the carnivorous Dinosaurs such as Megalosaurus and Tyrannosaurus were much smaller than the vegetarian Iguanodon, Diplodocus, Brontosaurus and Triceratops on which (or on the like of which) they preyed—just as a tiger is smaller than a buffalo, and a wolf smaller than a horse. It is owing to causes of this nature that the life of some animals, and consequently their growth, is limited in duration. Occasionally the common lobster lives to a great age, and grows to be more than 2 ft. long. But he is doomed by his size; the smaller lobsters "go quickly around" and get all the food (carrion of the sea), and the big fellow has to starve. The whale-bone whales, it is true, take animal food; but it occurs in the form of minute sea-slugs and shrimps, which fill the surface waters in countless millions over hundreds of miles of ocean. Hence the whales of this kind have only to swim along with their mouths open through an unlimited supply of luscious food. The size of terrestrial animals is also, it appears, definitely related to the natural water-supply. There are very few small quadrupeds in the interior of Africa. On account of frequent "drought," the mammals have often to run a hundred miles or more in search of water. Only animals as big as the larger antelopes and the zebra can cover the ground. The smaller kinds die (and have, in fact, died out in past ages) in these regions of sudden drought.

The gigantic reptile Diplodocus on land.

FOOTNOTE:

[4] The elephant, the thigh-bone of which, measuring nearly 3 ft. in length, is drawn in Fig. 32, is a large Indian one. This species is exceeded in size by the African. See "Science from an Easy Chair," Second series, p. 123.—The largest elephant the bones of which are known is the Elephas antiquus of the Pleistocene, bigger than either of the living species and bigger than the mammoth, Elephas primigenius. The arm-bone (humerus) of one

of this species (Elephas antiquus) lately dug up near Chatham and now in the Natural History Museum, is 4 ft. 3 in. in length.

CHAPTER VII

WHAT IS MEANT BY "A SPECIES"?

THOSE who take an interest in natural history must find it necessary to know what the naturalist means by "a species" of animal or plant. What does he mean when he says: "This is not the same species as that," or "This is a species closely allied to this other species," or "This is a new species"? What are the "species" concerning the origin of which Darwin propounded his great theory? There is really no English word which can be exactly used in place of the word "species." I often have to use the word when writing about plants or animals, and should like once for all to say what is meant by it. One might suppose that a "kind" is the same thing as a species. And so it often is; but, on the other hand, by the word "kind" we often mean a group including several species. For instance, we say the "cat-kind" or the "daisy-kind," meaning the "cat-like" animals or the "daisy-like" plants. The expression "the cat-kind" includes the common cat and the wild cat, and even leopards, lions, and tigers, each of which is a species of cat. And by the "daisy-kind" we understand a group including several species of daisies, such as the common daisy, the ox-eye daisy, the camomile daisy, the michaelmas daisy, and others. Hence we cannot translate species simply by the word "kind." "Kind" is the same word as "kin"—"a little more than kin and less than kind," runs Hamlet's bitter pun. "Kind" means a group held together by kinship, and it may be a larger or a smaller group held together by a close kinship or by a more distant one. "Sort," again, will not serve our purpose as an English translation of "species." For, although "a sort" implies a certain selection and similarity of the things included in the "sort," the amount of similarity implied may be very great or it may be indefinitely vague and remote. Hence naturalists have to stick to the word "species," and to use it with a clear definition of what they mean by it.

Suppose we get together a large unsorted collection— many hundred "specimens" or individuals—of the common butterflies of England. Then, if we look them over, we shall find that we can pick out and arrange the specimens into definite groups, according to their colour-pattern. We find that the kinds which we readily distinguish are called in English the swallow-tails, the whites, the sulphurs, the clouded yellows, the tortoise-shells, the peacocks, the red admirals, the painted ladies, the gatekeepers, the meadow browns, the heaths, the coppers, and the blues. There might be others in such a collection, but that is enough for our purpose. On examining the specimens closely, we find that the colour-markings and "venation" or network by which the wings are marked and the shape of the wings, body, and legs of all the specimens of the swallow-tails are almost exactly alike, and unlike those of any of the others. We shall find if we have a dozen or two specimens that there is a slight difference in the pattern, size, and colour of wing of some of the swallow-tails, dividing them into two groups, which we soon ascertain to be the males and females; but this is so small a difference that we may ignore it. The swallow-tail is obviously and at once distinguished from any of the other butterflies in the collection by its colour-pattern and shape. So also with the others, there will be many specimens in each case agreeing in colour and pattern, and recognizable and distinguishable from the rest by the colour-pattern and by the "venation" or "nervures" of the wings. If we collect butterflies again in other years and in other parts of the country, we find the same set of shapes and patterns exactly, corresponding to what we have learnt to call swallow-tails, whites, sulphurs, clouded yellows, tortoise-shells, etc. There are, we thus learn, several distinct, unchanging kinds of butterfly, which are common in this country, and appear every year. Similarly we may go into a meadow in spring, and gather a number of flowers, and a naturalist will roughly arrange our bouquet into "kinds"; there will be the buttercups, the daisies, the clovers, the dead nettles, the poppies, the roses, the orchids, etc.

If, now, we look more carefully at our collection of butterflies, sorted out roughly into kinds or species, we shall find that the "whites," although holding together by a close similarity in having merely white wings edged and spotted with black, yet differ amongst themselves, so that we distinguish a larger kind, the large garden-white, and a smaller, commoner kind, the smaller garden-white, and we distinguish also the green-veined white, and possibly the rare Bath white, each of them differing a little in their spots as well as their size. These different sorts of "whites" can, once our attention is drawn to the matter, be readily distinguished from one another, and constantly are found in our collections. We thus arrive at the conclusion that, though the whites are much alike, and are a kind distinct from the other kinds of butterflies, yet the "whites" themselves can be divided into and arranged as several kinds distinct from one another. In fact, we discover (and an illustrated book on butterflies confirms us in the conclusion) that there are several ultimate kinds of whites which cannot be further separated into groups. These are what are called "species." The whites are therefore not a single species, as are our British swallow-tails, but a group of species, closely related to one another. We find the same thing to be true with regard to the blues. Though they are much alike, agreeing in a variety of details of spotting and colour, yet we can distinguish the chalk-hill blue, the common blue, the azure-blue, the Adonis blue, and others, as distinct "species" of blues. Then, again, when we carefully examine our English specimens of tortoise-shells, we find that there are two distinct "species"—the greater and the smaller—differing not only in size, but in pattern; and when we compare with these the painted lady and the peacock and the red admiral, we find that there is a certain agreement of wing-pattern (venation and outline) and details of shape among them all, although their tints and the shape of the spots and bands of colour differ. These different species "hold together" just as the whites do and just as the blues do. Naturalists have met the need for expressing this similarity of a number of distinct species to one another by introducing the term "genus" for such a group. In fact we arrange several species into a "genus." The "genus" is a "kind," but a more comprehensive "kind," than is a species. The species is an assemblage of *individuals* closely alike to one another; the genus is a group of *species* which are more like to one another than any of them are to other species.

Naturalists give to every genus a name, and also a name to each species in the genus. Since we naturalists want to know what butterflies or other species of animals and plants are found in other countries, and to be sure that we all (whatever our native language may be) mean the same thing by a name, Latin names are given to the genera and the species, and are necessarily used when one wishes to be sure that one is understood. The greatest trouble is taken to make certain that the name used is applied only to the original species and the original genus to which it was applied, for only so can one be sure that a writer in America or one in Italy or France means the same thing by a name as we do here in England. This is rendered possible and is actually brought about by the preparation of catalogues in which the species are described and figured, especially with regard to obvious points of detail which are constant, and are called "specific characters." These are chosen for special description, not haphazard, but with a view to their being recognized with certainty by those who study other specimens. Another extremely important proceeding in connection with this purpose of uniform naming, which involves vast labour and expense, is the maintenance of great collections of preserved animals and plants by the State in all civilized countries. In these collections either the original specimens to which names were given by recognized describers (called "type-specimens" or "the type") are preserved, or else specimens which have been compared with those original described specimens, and authoritatively ascertained to be the same as the "type." The maintenance of accuracy and agreement in regard to the names of all the "species" of plants and animals is a big task. It is now carried out by international councils, in which the skilled naturalists of the world are represented. Certain principles have been agreed upon as to the method of determining the priority of one name over others which have been employed for one and the same species by naturalists of different countries and at different times, and a general agreement as to what names are to be used has been arrived at. It is a matter which has involved a great deal of uncertainty and dispute, and still causes difficulty. By the exercise of good sense, and in consequence of the existence of an urgent desire really to understand one another, there is now every year an increasing uniformity and agreement among naturalists about the exact name to be applied to every species of living thing.

Returning to our collections of butterflies and meadow flowers, we may take the names of some of the species and genera as an example of the system of naming in use by scientific naturalists. The common swallow-tail is assigned to the genus Papilio. Its "specific name" is "Machaon," given to it by Linnæus, hence it is spoken of as Papilio Machaon. It is found in various parts of Europe as well as in England. But in Central Europe (often seen in Switzerland) there is also another species of swallow-tail, which only occurs as a rare accident in England. This is the pale swallow-tail, differing, not only by its paler colour but by definite spots and markings of the wings, from the English species. Its species name, or "specific name," is "Podalirius," and so it is known as Papilio Podalirius. Species of Papilio are found all over the world; more than 500 are known. Our two commonest whites belong to the genus Pieris—they are called respectively Pieris brassicæ (the larger) and Pieris rapæ (the smaller). The green-veined white is Pieris napi. Each of these three is called after the plant, cabbage, rape, or turnip, on which its caterpillar feeds. The rare Bath white is Pieris daplidice. Its caterpillar feeds on mignonette. There are dozens of species in other parts of the world allied to our "whites," which naturalists have carefully distinguished and characterized by their marks.

Several of our most beautiful species of English butterflies which are much alike have been enrolled in one genus—the genus Vanessa. This genus includes the great tortoise-shell, called Vanessa polychloros; the smaller tortoise-shell, Vanessa urticæ; the peacock, Vanessa Io; the painted lady, Vanessa cardui; the red admiral, Vanessa Atalanta; and the comma butterfly, Vanessa C-album. There are other European, Asiatic, and American species of Vanessa.

In the same way we find with our meadow plants that what we at first thought was a single kind, "*the*" buttercup really bears a name applicable to a genus in which are several common species. The genus is called Ranunculus, and there are several common English species with yellow flowers, but distinguished from one another by definite characters. They are Ranunculus acris, Ranunculus flammula, Ranunculus bulbosus, Ranunculus arvensis, Ranunculus ficaria (the lesser celandine). And then there is the white-flowered Ranunculus aquatilis—a common pond plant. Clover, again, is by no means the name for a single species. The clovers form the genus Trifolium, and in any English meadow we may come across the white clover, Trifolium repens; the red clover, Trifolium pratense; the hop clover, Trifolium agrarium: the strawberry clover, Trifolium fragiferum; the haresfoot clover, Trifolium arvense. So it is with the plants which at first sight we distinguish merely as "daisies." There are several distinct genera of daisies—Aster, Bellis, Chrysanthemum (ox-eye), Anthemis (camomile), and others, with several distinct species in each genus.

Enough has been said to show the reader that the mere notion of "kinds" does not carry the same meaning as "species," but that there are a number of regularly occurring definite forms of both animals and plants which can be arranged in groups consisting only of individuals which are very nearly identical with one another. A group of living things of this degree of likeness is called "a species," and receives a name. A less degree of likeness holds together a number of species to form what we call a genus, and the name of the genus is cited together with the name of the species when we wish to speak of the species with clearness and certainty. This system of double names we owe to the great Swedish naturalist of the eighteenth century, Linnæus. He proposed also that the relationships of living things to one another should be further expressed by grouping like genera into "families," then like families into "orders," and like orders into "classes." And since his day we go further and group classes into "phyla" or great stems of the animal pedigree. In this way a complete hierarchy or system of less and more comprehensive groups has been established, and is the means by which we indicate the natural groups of the family-trees of plants and of animals, what, in fact, is called the "classification" of each of these great series of living things. Linnæus compared his system of groups to the subdivisions of two armies. Thus, the one army represents the whole animal series, the other the whole vegetable series. An army is divided into (1) "legions," these into (2) "divisions," "divisions" into (3) "regiments," regiments into (4) battalions, and battalions consist of (5) companies, consisting of individual soldiers. According to Linnæus, we may compare the legions to classes, which are divided into orders, comparable to divisions; these into families, comparable to regiments; these into genera, comparable to battalions; and these into species, comparable to companies, or ultimate groups of individual units or soldiers.

Just as the legions, divisions, regiments, battalions and companies of an army have each their own name or at any rate a distinctive numeral assigned to them in order that they may be cited and directed, so are names given to each class, order, family, genus and species of the classification or enumeration of the kinds of animals and plants. Here, for instance, are the names of the greater and smaller groups in which our common "white" finds itself enrolled. *Class*—Insects. *Order*—Lepidoptera. *Family*—Pieridæ. *Genus*—Pieris. *Species*—brassicæ.

CHAPTER VIII

MORE ABOUT SPECIES

I WROTE in the last chapter of the recognition of that degree of "likeness" or kinship in animals and plants which we point to by the word "species," and of the grouping of several similar species to form a "genus," and of several genera to form a family, of families to form orders, and of orders to form classes—and of the giving of names to all these groups. Whilst the making of this or that lot of species into a distinct *genus*, and giving it a new name is a mere matter of convenience for the indication of more or less important agreements and divergences, and is to a large extent arbitrary or an expression of opinion—it has always been recognized among naturalists that the group called "a species" is not a mere convention, but has a real natural limitation. It is true that the actual things which we see in studying natural history are so many units or individuals. But the possibility of arranging these by pattern, colour and shape into ultimate companies of which all the units are alike and differ from all the units of another company, has been regarded as a natural fact of primary importance and not a mere convention or convenience. The conception of the "naturalness" of a species depends really upon a further qualification of great importance as to what we naturalists understand by it.

We find by rearing plants from seed and by causing animals to breed under actual observation that the individuals of a species pair with one another, and not with individuals of other species, and further, that the young which they produce are like the parents—show themselves, in fact, to be of the same "species." The species continually year after year reproduces itself with little variation, though some variation does occur. The faculty of pairing only within the group, of never naturally breeding with members of other groups, has accordingly been adopted as a test of species. Hybrids between two species do not occur, except in very rare cases, in the state of nature. It is not always the case that the members of two species cannot possibly pair together, but it is the fact that they do not do so. Man sometimes brings about such crossing or hybridization, and it is a curious fact that the hybrids are often infertile or give rise only to weakly offspring, which could not survive in the natural struggle for existence. Sometimes, however, when the two hybridized species happen to come from regions of the world remote from one another, the resulting hybrids establish a vigorous race. There are real obstacles (of which I will say more below) in natural conditions to hybrid-breeding between any two species which occur naturally in the same territory. Thus the idea of a species is expanded so as to be not merely "a group of individuals of constant likeness in form and characteristics," but we add to that definition a living or constitutional quality expressed by the words, "which produce fertile offspring by pairing with one another, but do not pair with the members of other species."

This enables us to distinguish the conception of a "species" from that of a "variety" or a "race." We find occasionally peculiarly-marked examples of a species of plant or animal, or even local races of peculiar form; but we do not regard them as "distinct species" if we find that they breed as a rule with the ordinary members of the species. The decisive test is the breeding. If the variety is found not to breed with the regular species, but to keep apart and breed only with other individuals like itself, then we say, "This is no mere variety! It is a distinct species!" Unfortunately we have vast series of animals, insects, and others, from all parts of the world, collected and preserved in our museums, of which we know only the dead preserved specimens. So that we cannot be sure in doubtful cases whether a series of forms differing a little from the ordinary members of a species indicate distinct species, as defined and tested by breeding. We have in such a case to note the difference, and record it either as a variety or as a species by a guess at the probabilities one way or the other. Naturalists really *intend* by the word "species" to designate a form represented by numerous like individuals, which, in the present natural conditions of the region they inhabit, have attained a certain "stability" of distinctive form and character (not without some variability within definite limits) and constitute a more or less widely distributed population, the members of which inter-breed but do not produce offspring with other allied species.

A good case by which to exhibit further our conception of a species is that afforded by the species which are united in the genus Equus—the horse-genus. There are living at the present day several wild kinds of Equus—namely, the wild horse, or Tarpan, of the Gobi desert of Mongolia, called after the Russian explorer Przewalski; two kinds of Asiatic wild ass, called the Kiang and the Onegar; the African wild ass, and two or three kinds of zebra. There are, besides, many kinds of domesticated horses, ranging from the Shetland pony to the Flemish dray horse, and from the Shire horse to the Arab. Then there are many kinds of fossil extinct horses known, some of which clearly must be placed in the genus Equus with the living kinds; others which have to be separated into special genera (Hippidium, Onohippidium, etc.). Now, as to the living forms or form-kinds of the genus Equus—which are we to regard as true species, and which are only varieties and races of lower significance than species? The answer is clear enough in regard to several of them. The wild Mongolian horse and all the domesticated horses are varieties, races, or breeds of one species, judged not only by such marks as the possession of callosities on both the hind and the fore legs, but also by the test of breeding. They breed together and produce persisting races. But the asses and the zebras, though they will form mules with the horse, do not in a state of nature freely breed with it. When an ass or zebra is mated by man with the horse it will produce hybrids, called "Mules," but will not in "a state of nature" *establish* a hybrid race. The asses and the zebras are distinct from the horse, not only in markings and certain details of shape and hair, but in the fact that they cannot be fused into one race with him. There are no sufficient experiments on the aloofness of zebras and asses from one another in regard to breeding, although it seems that they cannot establish a mixed race, and are, therefore, distinct species judged by that test as well as by their form and marking. It is not known whether the so-called species of wild ass—the Asiatic and the African—would prove to produce fertile or infertile mules if intercrossed, nor has the test been applied to the very differently-marked local races of the African zebras—Grevy's zebra, Burchell's zebra, and the mountain zebra. It is likely enough that the three or more "species" distinguished among zebras on account of their being differently striped, and existing in different localities, would be found to breed freely together, and prove themselves thus to be entitled to be regarded as local "varieties" or "races," but not as fully-separated true species.

Thus one sees how difficult it is to have knowledge of the breeding test, even in regard to large animals. It is obvious that the difficulty of obtaining it in regard to the thousands of kinds of minute creatures is much greater. Yet when they say, "This is a distinct species," naturalists do mean that it is not only marked off from other animals or plants most like to it by a certain shape, colour, or other quality or qualities, but that it breeds apart with its own kind and does not naturally hybridize with those other forms most like to it.

Although the kind of naturalist called a "systematist" who makes it his business to accurately describe and record and distinguish from one another all the existing species of some one group—say, of antelopes, of mice, of flowering plants, of fishes, or of fleas—has only a knowledge in a few instances of the breeding of the organisms which he describes as "distinct species," he yet does know, in regard to some one or more of his species in most groups, the facts of pairing and reproduction, and what are the limits of variation in the markings and other characteristics of at least one or two species definitely submitted to the "breeding test," that is to say, ascertained to be "true physiological species," kept apart by deep-seated chemical differences in their blood and tissues. Hence it is legitimate for him, by careful balancing and consideration of all the facts, to determine—not absolutely, but by analogy—the value to be assigned (whether as indicating true species or merely varieties capable of pairing with the main stock) to points of difference among the specimens of a dead collection brought from some distant land or from some position in which it would be impossible to make observations with regard to "pairing" and "breeding true."

Some 400 species of fleas have been described, and we are certain as to the value of the characters relied on to distinguish those species, owing to what we know of the breeding of some common species of fleas. The flea of the domestic fowl, that of the domestic pigeon, that from the house-martin, and that from the sand-martin—used to be considered as one species until they were carefully examined twenty years ago. In reality each of them has its own peculiar "marks," and they do not mix with one another. The nests of the sand-martin yield only one species of flea, namely that peculiar to the sand-martin. The hen-house, the dove-cote, and the nests of the house-martin yield each their flea maggots, which can be reared and become in each case a distinct species with definite recognizable "characters." On the other hand, the flea of the rabbit gives an opportunity of studying the limits of variation in a "good" species. Rabbit warrens swarm with the rabbit flea, and often a great number are found on one rabbit, the individual fleas "varying"—"differing" from one another to a slight extent. The "systematist" thus gets to know what organs are variable within the limits of an undoubted physiological species of flea, and what are comparatively constant—so that he can form a reasonable opinion about the claim of other specimens which he may receive without full history of their habits, to be regarded as true distinct species.

The fact that most important chemical differences of the blood and digestive juices often accompany the small external differences which enable us to distinguish one species of animal or plant from another, makes it obvious that the knowledge of species is a very valuable and necessary thing. One species of flea, the Pulex Cheopis, habitually carries the plague bacillus from animals to man, and is a cause of death; other species, extremely like it in appearance, but distinguishable by a trained observer, do not carry the plague bacillus, but if they swallow it, destroy it by digestion. One species of gnat, the common grey gnat, digests and destroys malaria germs when it sucks them up with blood; in an allied species, the spot-winged gnat or Anopheles, the chemical juices of the gut allow the germ to live in it and multiply, and so to be carried to men by the gnat's bite. So with many other flies and parasites the recognition of the dangerous species is of vital importance, and that recognition often depends on minute features of form and colour not at once obvious to an ordinary observer.

But this recognition of distinct species is, from the point of view of the study of Nature, only a preliminary to the question, "How did these species come about? How is it that there are so many species, some very like one another, forming genera, and these genera grouped into related families, these into larger groups, and so on, like the branches of a family tree?" The answer to these questions given by Linnæus was: "There are just so many species as the Infinite Being created at the beginning of things, and they have continued to propagate themselves unchanged ever since." The answer which we give to-day is that the appearance of a huge family tree which our classification of animals takes is due to the simple fact that it really is neither more nor less than a family tree or pedigree—the "tree of life," of which the green leaves and buds are the existing species. Further, we hold that the existing species of a genus have "come into existence" by natural birth from one ancestral species, its offspring having slightly varied (we are all familiar with this individual variation in our own species, in dogs, cats, trees, and shrubs), and that the varieties have wandered apart and become continuously emphasized and selected for survival by their fitness or suitability to the changed conditions around each of them. Meanwhile a natural destruction, or failure of intermediate forms to survive, has gone on.

CHAPTER IX

SPECIES IN THE MAKING

A SERIES of important conceptions are implied in the word "species," as used by naturalists. Some of these we have noted in the last chapter. There is first, as a starting-point, the conception that a species is a number or company of individuals, all closely and clearly alike (though presenting some minor individual variations), and capable of sharp separation by certain "characters" from other similar groups or companies. Then follows the addition (2) that the species is constant if the conditions of life are not changed, or but little changed, and that year after year it reproduces itself without change. It has a certain stability (but not permanent immutability) greater in some species than in others. Next we find (3) that the species constitutes a group of individuals which have descended by natural breeding from common parents, not differing greatly from the present individuals. They are, in fact, one "stock." Then (4) that the species is a group, the individuals of which pair with one another in breeding, but do not pair with the individuals of another species, and that this is due to various peculiar and inherent chemical, physiological and (in higher animals) psychological characteristics of the species.

We have now further to note that species have their special geographical *centres of origin* from which most spread only a small distance, whilst others have a wonderful power of dispersal, and have become cosmopolitan. Moreover, we find that some species are numerically very abundant, others very rare; that rare and abundant species have often invaded each other's territory, and exist side by side.

Whilst we often find a number of species, fifty or more, so much alike that we unite them in a single genus (as, for instance, in the case of the cats, lions, tigers, leopards, which form the genus "Felis," and the hundred or more species of the hedge brambles or blackberries, which form the genus "Rubus"), there are many species which to-day have, as it were, lost all their relatives and stand alone, the solitary species in a well-marked genus, or have perhaps only one other living co-species. And sometimes (curiously enough) that one co-species is an inhabitant of a region very remote from that inhabited by the other. Thus the two living mammals called tapirs (genus Tapirus) inhabit, the one the Malay region, and the other Central America. This is explained by the fact that tapirs formerly existed all over the land-surfaces of North Europe, North Asia, and North America, which connect these widely-separate spots. We find the bones and teeth of the extinct tapirs embedded in the Tertiary deposits of the connecting regions.

Once we have gained the fundamental conceptions as to what is meant by a "species," we are able intelligently to consider innumerable facts of the most diverse kind as to their peculiar structure and colours, their number, localities, their interaction and dependence on other living things, their modifications for special modes of life, their isolation or their ubiquity. We can discuss their genetic relations to one another, and to extinct fossil species, which have all been to a very large extent "accounted for" or "explained" by Mr. Darwin's theory of the origin of species by the natural selection of favoured races in the struggle for existence. But there is always more to be made out—difficulties to be removed, new instances to be studied. The classification of the genera of plants and animals, with their included species into larger groups, helps us to state and to remember their actual build and structure, and to survey, as it were, the living world, from the animalcule to the man, or from the microbe to the magnolia tree. Every one interested in natural history should carry in his mind as complete a scheme of the classification of animals and plants as possible.

The older naturalists held that species were suddenly "created" as they exist, and have propagated their like ever since. Darwin has taught us that the present "species" have developed by a slow process of transformation from preceding species, and these from other predecessors, and so on to the remotest geologic ages and the dawn of life. The agents at work have been "variation"—that is to say, the response to the never-ceasing variation of the surrounding world or environment—and the survival in the struggle for existence of the fittest varieties so produced.

There is nothing surprising or extraordinary in the existence of variation. The conditions of life and growth are never absolutely identical in two individuals, and the wonder is not that species vary, but that they vary so little. The living substance of animals and plants is an extremely complex chemical substance, ever decomposing and ever being renewed. It is the most "labile" as it is by far the most elaborately built-up chemical body which chemists have ever ventured to imagine. It differs, chemically, not only in every species but in every individual and is incessantly acted upon—influenced as we may say—by the ever-changing physical and chemical conditions around it. At the same time it has, subject to the permanence of essential conditions, a definite stability and limitation to its change or variation in response to variations of its environment. That part of the living substance which in all but the lowest plants and animals is set aside during growth to form the eggs and sperms by which they multiply or "reproduce" themselves, is called the "germ-plasm," and is peculiarly sensitive to variations in (that is a *change* in) the environment of the plant or animal.

New conditions of life (locality and climate)—unusual food or reproductive activity—act often in a powerful way upon the germ-plasm and cause it to vary—that is to say, they alter some of its qualities, though not necessarily disturbing in any way the general living substance of the organism so far as to produce any important change perceptible to the human eye. In consequence, the young produced after such disturbance of the germ-plasm are found to differ more from their parents than in cases where no such disturbance has been set up by the natural never-ceasing variation of the surrounding world. This fact is well known to horticulturists and breeders, and is made use of by them. When a gardener wishes to obtain variations of a plant from which to select and establish a new breed, he deliberately sets to work to disturb—to shake up, to act upon in a tentative, experimental way—the germ-plasm of one or more parent plants by change of soil, climate, food and often by cross-fertilizing them with another breed or variety. In this way he to some extent "breaks" the constitutional stability of the germ-plasm of the plant and obtains abundant "variations" in the offspring. These are not precisely foreseen, and show themselves in all parts of the new generation. But some of them are what the gardener wants, and are "selected" by him for retention, rearing and breeding.

The response of the germ-plasm of organisms to the stimulus of new environmental conditions has been compared to that of the well-known pattern-producing toy—the kaleidoscope. The bits of glass, beads and silk which you see in a kaleidoscope, forming by reflection in its mirrors a beautiful and definite pattern, are changed by a simple vibration caused by tapping the instrument into a very different pattern, the coloured fragments being displaced and rearranged. The apparent change or variation is very great though produced by slight mechanical disturbance, and the new pattern is altogether without any special significance—the fortuitous outcome of a small displacement of the constituent coloured fragments. We can imagine that similarly slight disturbances of the organic molecules of the germ-plasm may produce considerable and important variations in it and the new growth to which it gives rise: and, further, that these variations may prove to be either (1) injurious, or (2) of life-saving value, or often enough (3) of no consequence whatever although bulking largely in our human eyes and thereby misleading our judgment of them.

There is no reason to doubt that the same sequence of events occurs in nature apart from man's interference. Changes occur in the earth's surface, or the organism is transported by currents of water or air into new conditions. The germ-plasm is "disturbed," "shaken" or "shocked" by those new conditions, and a variation, in several structures and qualities of the offspring subsequently produced, follows. Then also follows the selection of one of the new varieties by survival of the fitter to the new conditions into which the organism has been transported or have developed in the region where it was previously established.

This process of germ-variation is obviously as necessary and constant a feature of the living organism as is the variation in the contour of land and sea and in the extent of the polar ice-cap—a necessary feature of the physical conditions of the terrestrial globe. But it is the fashion with a certain school of writers nowadays to declare that "variation" in organisms is a "mystery" unsolved. Another very common and almost universal error is to overlook the fact that variation is constitutional and affects whole systems of organs and their deeply related parts, and is *not*, as it is so frequently and erroneously assumed to be, a mere local affair of patches and scraps visible on this or that part of the surface of an animal or plant. These superficial "marks," readily seen and noted by the collector, are rarely of any life-saving importance: they are but the outward and visible signs of deep-lying physiological or constitutional change or variation. The varying organism has, like Hamlet, "that within which passeth show" and the superficial variations (like his "inky cloak" and other customary features of mourning) are but "the trappings and the suits" of a deep-lying change. Variation is not an inexplicable mystery, nor, on the other hand, are "varieties" sufficiently dealt with and their nature appreciated when one or two surface peculiarities are enumerated by which the collector can recognize them. A deeper study of the varying organism is both possible and needed.

If the gradual formation of new species from ancestral species is a true account of the matter, we must expect to find, at any rate here and there, if not frequently, traces of the process—for instance, gradations, or series of intermediate forms, connecting new, well-established species with the ancestral form or with one another. We do find such gradations—sometimes more, sometimes less, completely persisting over a wide tract of country, or discoverable in the fossiliferous deposits containing the remains of extinct animals.

For instance, when we look at the butterflies of a much larger region than our little island—namely, at those of a great continent like Africa or South America—we find that there are species which show gradations. Thus at a series of points, A, B, C, D, separated by some hundreds of miles from each other, we find a corresponding series of butterflies which are apparently closely similar species of one genus, differing by a few spots of colour, or darker and lighter tint, much as our Large White, Garden White, and Green-veined White differ. But when the butterflies are caught which occur at points intermediate between A and B, B and C, C and D, we find intermediate varieties, and, in fact, if we get a very large number from intermediate regions, we can, in some instances, arrange them in line so that they constitute a graduated series of forms, each being scarcely distinguishable from the one before or the one behind it, yet differing clearly from one a dozen places away. In such cases there is often evidence to show that the variety found at A breeds with that found at B, that of B with that of C, of C with D, so that they form an inter-breeding group, though perhaps the varieties at D will not pair with those at A, or even with those at B. Then sometimes we find in such a series, otherwise complete, a gap. Let us suppose it is between the butterflies of B and C. We find the series of gradations nearly complete, but some natural condition—such as the encroachment of the sea, or the slow elevation of a mountain range, or the climatic destruction of the necessary food-plant—has "wiped out" a few forms somewhere between those of B and C. They no longer exist. The series is no longer connected by inter-breeding forms; those occurring from A to B and some distance beyond are one "species" varying in the direction of the series C to D, but abruptly broken off from the latter. The series C to D is also a "species" with graduated varieties, but distinct; it is cut off from the lot once in continuity with it by the destruction of the intermediate forms inhabiting an intermediate area. Thus the one species becomes two, and these may again break up, and, having become thus disconnected and stabilized, they may spread over one another's territory—fly side by side and yet remain distinct forms which do not pair together—although originally they were varieties spreading from a common centre, where the ancestral species lived and multiplied.

Other similar gradational series of an interesting character have been noticed in the case of fresh-water fossil snail-shells. In the layers of clay and marl exposed by digging a railway cutting or a pit we may find that the successive layers represent a continuous deposit of 100,000 years or more, and we find sometimes that a form of snail-shell (not a species living to-day) occurs in the lowest stratum very different from that occurring in the highest stratum—the lowest being short and spherical, the highest elongated and of differing texture. In the intermediate layers, each 6 or 12 ins. thick and occupying perhaps altogether 30 ft. of vertical thickness, we find a graduated series of snail-shells leading almost imperceptibly from the oldest lowest form to the latest uppermost form. Such cases are known. But it is an exceptional thing to find these graduated series either spread over an area of the earth's surface, or following one another in successive strata. When they came into existence they were rapidly superseded and destroyed as a rule, and have left only one or two widely-separated examples of the intermediate forms. This we should naturally expect by analogy from what we know of the successive traces of human manufactures in the deposits on the site of some of the great cities of the ancient world which have been carefully excavated layer by layer. But still we have the important fact that here and there such gradational series have been found, and we are justified in considering a few isolated intermediate forms (which often occur connecting two greatly-differing species) as survivors of a former complete graduated series of intermediate forms, which came into existence by slow modification of an ancestral stock, and may, when the stock was widely spread over a continental area, not merely have succeeded one another in time, but actually coexisted in neighbouring regions.

There are many remarkable facts bearing upon the origin of "species," the description of which fills volumes written by such men as Darwin, Wallace, Poulton, and others, and become interesting to every one who has gained a correct notion of what naturalists mean by a "species." I will cite one in order to illustrate this. The bird which we call the red grouse, or nowadays simply "grouse" (the old Scotch name for it was "muir-fowl"), is one of twenty-four birds (among the 400 species of birds which live in the British Islands), including several kinds of titmouse, the goldfinch, bullfinch, song-thrush, stonechat, jay, dipper, and others which are very closely similar to species of birds living in Continental Europe, yet show some definite and constant marks, such as small differences in the colour of a group of feathers, enabling us to distinguish the British from the Continental forms. Are these twenty-four British forms to be regarded as distinct species?

The red grouse is placed in a genus called "Lagopus," of which there are several species in the northern hemisphere. In Scotland the red grouse, which is distinguished as Lagopus Scoticus, is accompanied by a rarer species of Lagopus, which lives in high, bare regions. This is the bird called by the Celtic name "ptarmigan"; it differs in several points from the red grouse, and acquires white plumage in the winter, which the latter bird does not; it is called Lagopus mutus. Now in Norway we find also two species of grouse or Lagopus, called "rypé" (pronounced "reeper") by the Norwegians. One is the same bird in every respect as the Scotch ptarmigan, and is known as "the mountain rypé." The other is very close to our red grouse, and is called "the common or bush rypé," and by English naturalists the "willow grouse," and by ornithologists "Lagopus salicetus." It agrees in habits, voice, eggs, and anatomical detail with our red grouse, but the back of the cock-bird of the red grouse and the whole plumage of the hen-bird have a darker colour. Moreover, the willow grouse, like the ptarmigan or mountain rypé, turns white— acquires a white plumage—in the winter which the red grouse does not. Are the red grouse and the willow grouse to be regarded as distinct species? Our British red grouse lives on heather-grown moors; the willow grouse prefers the shrubby growths of berry-bearing plants interspersed with willows, whence its name. Their food differs accordingly. Formerly the red grouse lived on the moors of the South of England, and when in Pleistocene times England was a part of the Continent of Europe the willow grouse and the red grouse were one undivided species inhabiting all the north-west of Europe. It is probable, though the experiment would be almost impossible to carry out, that were the eggs of a number of willow grouse now brought to Scotland and hatched on the moors, they would tend to keep apart from the native red grouse, and not inter-breed with them, in which case we should say that the Scotch form is a "species on the make," or, even, a completed and distinct species. On the other hand, it is possible that the two forms would freely pair with another, and that the colour and winter coat of the one (probably that of the Scotch form if the experiment were tried in Scotland) would predominate, and after some generations no trace of the other strain would be observable.

CHAPTER X

SOME SPECIFIC CHARACTERS

AN interesting case, showing that qualities which are life-preserving under certain severe conditions exist in some varieties of a species and not in others, was recorded some eight years ago. After a very severe "blizzard" 136 common sparrows were found benumbed on the ground by Professor Bumpus at Providence, United States. They were brought into a warm room and laid on the floor. After a short time seventy-two revived and sixty-four perished. They were compared to see if the survivors were distinguished by any measurable character from those which died. It was found that the survivors were smaller birds (the sexes and young birds being separately compared) than those which died, and were lighter in weight by one-twenty-fifth than the latter. Also, the birds which survived had a decidedly longer breastbone than those which died.

Similarly, the late Professor Weldon found that in the young of the common shore-crab, taken in certain parts of Plymouth harbour, those with a little peculiarity in the shape of the front of the shell survived when those without this peculiarity died. Many thousands were collected and measured in this experiment. It is not necessary to suppose that the distinguishing mark of the survivors in such cases is "the cause" of their survival. Such marks as the breadth of the front part of the crab's shell and the length of a bird's breastbone very probably are but "the outward and visible signs of an inward and (physiological) grace."

The marks, little peculiarities of colour and proportionate size, or some peculiar knob or horn, by which the student of species distinguishes one constant form from another, can rarely, if ever, be shown to have in themselves an active value in aiding or saving the life of the species of plant or animal. The mark or "character" is an accompaniment of a chemical, nutritional, physiological condition, and is in itself of no account. It is what is called "a correlated character." Such, for instance, is the black colour of the skin of pigs which in Virginia, U.S., are found, as stated by Darwin, not to be poisoned by a marsh plant ("the paint-root," Lachnanthes tinctoria), whilst all other coloured and colourless pigs are. The pigs which are not black develop a disease of their hoofs which rot and fall off, causing their death when they eat this special plant "the paint-root." The colour does not save the pig—it cannot correctly be called the *cause* of the pig's survival—but is an accompaniment of the physiological quality which enables the pig to resist the poisonous herb. So, too, with white-spotted animals. They are known to breeders as being liable to diseases from which others are free. Fantail pigeons have extra vertebræ in their tails, and pouter pigeons have their vertebræ increased in number and size. But the vertebræ were never thought of and "selected" by the breeders. They only wanted a fanlike set of tail feathers in the one case, and a longer body in the other. Some varieties of feathering maintained by pigeon breeders lead to the growth of abundant feathers on the legs (as in Cochin-China fowls), and it is found that these feather-legged pigeons always have the two outer toes connected by a web of skin. If it were a stabilized wild form we should separate it as a species on account of its webbed toes, yet the real selection and survival in the hands of the breeder had nothing to do with the toes or their web, but was simply "caused" by these pigeons having feathers of "survival or selection value" in his judgment. Male white cats with blue eyes are deaf. If deafness were ever an advantage (a difficult thing to imagine), you would get a species of cat with white hair and blue eyes, and be led to distinguish the species by those characters, not by the real cause of survival, namely, deafness. Not enough is yet known of this curious and very important subject of correlation, but its bearing on the significance of "specific characters" is sufficiently indicated by what I have said.

An interesting group of species, three of which are to be purchased alive through London fishmongers, are the European crayfishes, not to be confused with the rock-lobster or Langouste (Palinurus), sometimes called "crawfish" in London, nor with the Dublin prawn (Nephrops). The little river crayfishes are like small lobsters, and were placed by older naturalists in one genus with the lobsters. Now we keep the European species of crayfishes as the genus Astacus, and the common lobster and the American lobster have been put (by H. Milne-Edwards) into a separate genus (Homarus). You can buy in London the "écrevisses à pattes rouges" of French and German rivers, which is called Astacus fluviatilis, and differs from that of the Thames and other English and European rivers (which you can also buy) called A. pallipes ("pattes blanches" of the French), by the bright orange-red tips of its legs, and by having the side teeth of the horn or beak at the front of the head larger and more distinct. The English crayfish grows to be nearly as large as the "pattes rouges" in the Avon at Salisbury, though it has nearly disappeared about Oxford. You can also sometimes buy in London the big, long-clawed Astacus leptodactylus of East Europe. There are two or three other species, named and distinguished, which do not come into the London market.

Crayfishes, lobsters and the like have groups of plume-like gills (corresponding in the most ancient forms to the number of the legs and jaw-legs) overhung and hidden by the sides of the great shield or "head" of the animal. The common lobsters and crayfishes retain most of these in full size and activity, but have lost in the course of geologic ages the original complete number. These plume-like gills—each half an inch or so in length—are attached, some to the bases of the legs and some to the sides of the body above the legs. In the ancestral form there were thirty-two plumes on each side, twenty-four attached to the bases of the legs, and eight placed each at some distance above the connection of one of the eight legs with the side of the body. It is those on the side of the body which have suffered most diminution in the course of the development of modern crayfishes (and lobsters) from the ancestral form provided with the full equipment of thirty-two gill-plumes on each side. In fact, only one *well-grown* gill-plume, out of the eight which should exist on each side of the body-wall, is to be found—and that is the one placed above the insertion of the hindermost or eighth of the eight legs (eight when we reckon the three jaw-legs as "legs" as well as the five walking-legs). In front of this the side or wall of the body is bare of gill-plumes though they are present in full size on the basal part of most of the legs. Nevertheless, when one examines carefully with a lens the bare side of the body overhung by the head-shield or "carapace," one finds in a specimen of the common English "pale-footed crayfish" a very minute gill-plume high above the articulation of the seventh leg and another above the articulation of the sixth leg. They are small dwindled things, as though on the way to extinction, and are the mere vestiges of what were once well-grown gill-plumes, and still are so in the rock lobster and some prawns. In the red-footed crayfish of the Continent (Astacus fluviatilis) yet another minute vestige of a gill-plume is found, farther in front, on the body-wall above the fifth leg on each side of the animal. This furnishes a definite mark or character by which we can distinguish the red-footed crayfish from the common English pale-footed one. But these three rudimentary gill-plumes in the red-foot species, and two in the pale-foot species are all that until lately were recorded. The region of the body-wall above the fourth, third, second, and first of the legs was declared to be devoid even of a vestige of the branchial plumes which were there in ancestral forms, and have been retained more or less in some exceptional prawn-like creatures allied to the crayfish.

Zoologists take a special interest in the crayfish because it is found to be a most convenient type for the purpose of teaching the principles of zoology to young students, and with that end in view was made the subject of a very beautiful little book by the great teacher Huxley. The conclusions above stated in regard to the gills are set forth in that book with admirable illustrative drawings, and the striking fact of the dwindling and suppression of the various gill-plumes is clearly explained.

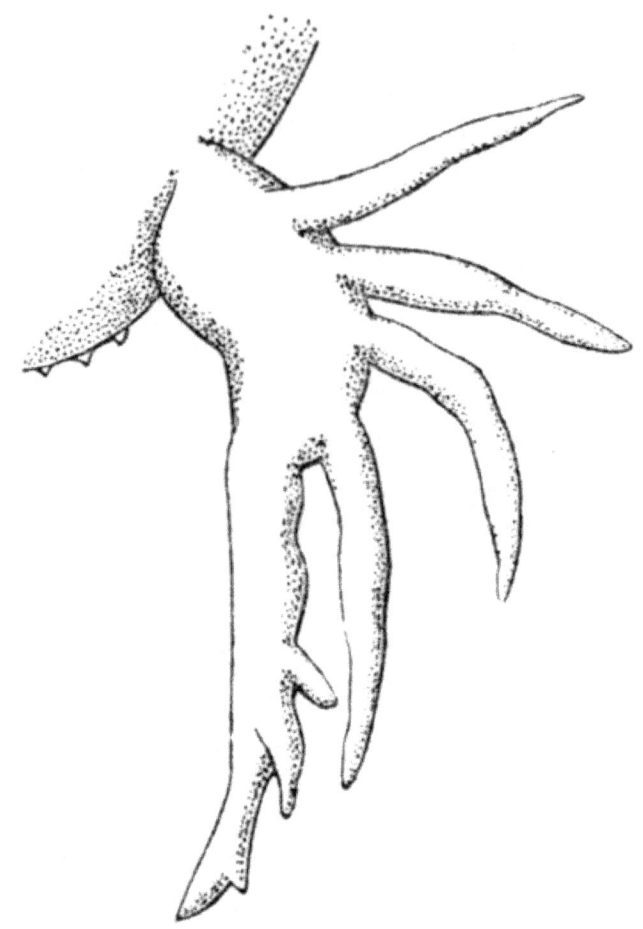

FIG. 33.—The rudimentary gill-plume of a crayfish from that part of the body-wall to which the first pair of jaw-legs (maxillipedes) is articulated. Found in the red-footed crayfish (Astacus fluviatilis) but in no other species of Astacus. It is one-fifteenth of an inch long. Drawn by Miss Margery Moseley in 1904. ("Quart. Journal of Microscopical Science," vol. 26 (1904-5).)

And now we come to an interesting discovery in this matter of the gill-plumes of crayfishes. Some fifteen years ago the daughter of my friend and colleague—Professor Moseley—was a member of the class of Elementary Biology at Oxford. She had to examine and identify these and other points in the structure of the crayfish. The class was supplied with specimens of the French red-footed crayfish "Astacus fluviatilis," as it is more readily obtained from fishmongers than our own "pale-foot" or "Astacus pallipes." She found in her specimen far forward on each side of the "head" a very minute gill far away from the others and previously unknown. The demonstrator in charge of the class refused even to look at her discovery. So she confirmed it by examining three other specimens—made drawings of the tiny branched gill (as shown in Fig. 33) and their position, and sent them to me in London. It was at once clear that she had discovered in this much studied little animal a very interesting pair of gills (right and left)—unknown to Huxley and the rest of the zoological world. She proceeded to examine specimens of A. fluviatilis from various rivers of Germany and France and always found the new gill-plume. She also showed (I supplied her with specimens at the Natural History Museum) that it was, on the other hand, absent from A. leptodactylus, A. pallipes, and all the foreign species (some from Asia) which are known, and she published an illustrated account of it in the "Quarterly Journal of Microscopical Science." This tiny gill-plume is placed very far forward on each side of the body, the farthest point forward at which any gill-plume is found in any kind of prawn, shrimp or lobster, namely in the region where the first pair of jaw-legs is attached, so that there are three empty spaces between it and the rudimentary gill over the fifth pair of legs, already known in the red-footed crayfish. It is only two millimetres long—about one-fifteenth of an inch! But its presence serves very distinctly to separate the red-footed crayfish, Astacus fluviatilis of French and German rivers, thus discovered to have four pairs of rudimentary gill-plumes, from the Astacus leptodactylus of the Danube basin and East Europe, which has only three pairs, and still more to emphasize the difference between it and our British species, the "white-foot" or Astacus pallipes, which has only two!

This little history is noteworthy, firstly, because it shows that a young student may, to use an appropriate term, "wipe the eye" of an expert observer and rightly venerated teacher (who would have delighted in the little discovery had he been alive), as well as the eyes of tens of thousands of students and teachers (including myself) who have studied the red-foot crayfish year after year, and missed the little gill. It is also interesting as showing us a good sample of a specific mark or character which has no survival value; that is, could not advantage the crayfish in the struggle for life. The fact is, that this one particular very minute forward pair of gill-plumes is like the other rudimentary gills—a survival in a reduced condition of a pair of gill-plumes which were well-grown, useful plumes aerating the blood, in the prawn-like ancestors of all crayfishes, lobsters, shrimps, and prawns, and is, owing to circumstances of nutrition and growth which we know nothing about but can vaguely imagine, retained by the red-foot species of crayfish, but lost by all other crayfishes, lobsters, common prawns and shrimps, and, in fact, only retained besides by a very few out-of-the-way kinds of marine prawns. That is the sort of thing which frequently has to serve as "a specific character" or mark, distinguishing one "species" from another.

A more ample discussion of the origin of species is not within the scope of this book. But I may say that until recently the conception that *every* organ, part and feature of a plant and animal *must* be explained, and can *only* be explained, as being of life-saving value to its possessor, and accordingly "selected" and preserved in the struggle for existence, was held by many "Darwinians" in too uncompromising a spirit. This conception was, really from the first, qualified by the admission that the life-saving value and consequent preservation of a structure must undoubtedly in some cases have been in operation in ancestors of the existing species, and is no longer operative in their descendants although they inherit the structure which has now become useless. Moreover, the operation of those subtle laws of nutrition and of form which are spoken of as the "correlation of parts in growth and in variation" (mentioned on p. 119) was pointed out by Darwin himself as probably accounting for many remarkable growths, structures and colour-marks which we cannot imagine to be now, or ever to have been in past ancestry, of a life-saving value. Nevertheless, the old "teleology," according to which, in pre-Darwinian days, it was held that every part and feature of an animal or plant has been specially created to fulfil a definite pre-ordained function or useful purpose, still influenced the minds of many naturalists. Natural selection and survival of the fittest were reconciled with the old teleological scheme, and it was held that we must as good Darwinians account for every structure and distinctive feature in every animal and plant as due to its life-saving value. Herbert Spencer's term, "the survival of the *fittest*," conduced to the diffusion of this extreme view: Darwin's equivalent term, "the preservation of *favoured* races," did not raise the question of greater or less fitness.

The extreme view is now, however, giving place to the recognition of the fact that the actual tendencies to variation—accumulated in the living substance of the various stocks or lines of descent and handed on during an immense succession of ages of change by hereditary transmission—counts for more in the production of new species and strange, divergent, even grotesque forms of both animals and plants than had been supposed.

Undoubtedly selection or survival of the fittest mainly accounts for the colouring and adaptive shaping of living things, and so for those several great types of modelling which arrest the eye and have excited the interest of inquisitive man. But there seems to be no justification for the assumption that in all cases a variation—that is to say, an increase or a diminution of the volume of some existing structure in proportion to other coexisting structures in the body of a living plant or animal—must be *either* favourable, that is, conducive to survival, *or* injurious, that is, tending to the defeat and destruction of its possessors or their race. On the contrary, it is the fact that there are vast areas and conditions related to countless myriads of living creatures in which variations of those creatures of large and imposing kind and degree are neither advantageous nor disadvantageous, but matters of *absolute indifference*, that is to say, without any effect upon the preservation or survival of their race or stock. Nature is far more tolerant than some of us were inclined to assume. In certain restricted conditions of competition and in regard to some special structures and components which are often so minute and obscure as to be not yet detected by that recent arrival, the investigating biologist—though sometimes, fortunately for him, large enough to jump to his eyes—it is undeniable that there must be a "survival" or "favouring" of individuals presenting a variation in increase, or it may be decreased, of this or that special feature of its "make-up" or structural components. But it is a more correct statement of the case to say that natural selection or survival preserves *not the fittest*, but *the least fit possible under the circumstances*—namely, all those which, however great their divagations and eccentricities of variation in other respects, yet at the same time attain to a minimum standard of qualification in those structures (or inner chemical qualities) essential for success in the competition for safety, food and mating determined by the particular conditions in which the competition is taking place. Consequently forms which are meaningless so far as standards of utility or "life-saving" are concerned, and are rightly described as grotesque, monstrous, gigantic or dwarfed—excessive (as compared with more familiar kinds) in hypertrophy or atrophy of their colouring and clothing, or of out-growths such as leaves of plants and limbs, jaws or other regions of the body of animals—are found existing in various degrees of eccentricity in every class of both plants and animals. Among animals such tolerated "exuberances of non-significant growth" are more striking than in plants. The group of fishes seems to be especially privileged in this way. They are freely variable in the position of the fins, the suppression or exaggeration of them, as well as of the scales on the surface of the body (*e.g.* leather carp and mirror carp). Take, for example, the mackerel and the salmon as standards of utilitarian adaptation of the body to an active life in sea or river, and then compare with theirs the astounding proportions of the sun-fish (Orthagoriscus) like a cherub "all head and no body," or the almost incredible Pteraclis—with its little body framed immovably between a huge dorsal and a huge ventral fin (see figures on p. 130). The fin-like crest of enormous size on the back of the great extinct lizard Dimetrodon of the Permian age supported by long bony spines is a similarly excessive and useless outgrowth. (This astonishing creature is shown in our Frontispiece.) Such exuberant products may be ascribed to an unrestrained "momentum" of growth which once set going by fortuitous variation has been *tolerated* but not *favoured* by natural selection. Or (as supposed by some) their excessive development may be due to the *persistence* of some nutritional condition which at first resulted in a moderate growth of the fin-like crests in question as a serviceable structure, but has persisted and increased long after the fin or crest has attained a sufficient size—simply because its increase though of no life-saving value—yet was not harmful and so did not bring its owner under the guillotine of natural selection. Such disproportionate exuberance of growth due to innate variability, tolerated but not specially favoured by natural selection, will account for many strange and grotesque forms of living things. From time to time in the long process of change, such exuberances may suddenly become of service and be, so to speak, taken in hand by natural selection, or they may become dangerous and lead to the extermination of the stock in which they have been previously tolerated.

Before my reader turns—as I hope he or she will do—to some handbook of zoology in which the genealogical tree or classification of the species of animals and of plants is treated at length, I will endeavour to give some estimate of the immense numbers of "species" which exist. As to mere individuals, it is impossible to form any estimate, but when we reckon up the teaming population of a meadow or forest in England, the hundreds of thousands of plants, including the smallest mosses and grasses, as well as the larger flowers, shrubs, and trees, the still greater number of insects, spiders, snails, and larger animals and birds, feeding on and hiding among them, and when we remember that in the ever-warm tropical regions of the earth life is ten or twenty times more exuberant than with us,—then the immensity of the living population of the land and water of the globe becomes as difficult to realize as are the figures in which the astronomer tells of the number and distances of the stars. On the other hand, some idea of the number of distinct species of animals and plants which have up to this date been recognized and described by naturalists as at present existing, may be formed by a statement of those which have been described in some of the more familiar groups. About 10,000 species of mammals have been described; about 14,000 of birds; 7000 of reptiles; 15,000 of fishes; 500,000 of six-legged insects; 14,000 of crustacea (shrimps, lobsters, crabs); 62,000 of molluscs (snails, mussels, etc.); 15,000 of star-fishes and sea-urchins; 5000 of corals and polyps; 3000 of sponges; and 6000 of microscopic protozoa. In all about 800,000 species of animals have been recorded, and probably as many more remain yet to be recognized and described.

The total number of described species of plants has never been estimated, but some idea of it may be formed from the fact that 1860 species of flowering plants alone have been distinguished in Britain, 17,000 in British India, and 22,000 in Brazil, not to mention those of Africa and Australia! These figures do not include the vast numbers of flowerless plants, the ferns, mosses, sea-weeds, mushrooms, moulds, lichens, and microscopic plants.

And then we have to add to these enumerations of living species the extinct species of successive geological ages, the remains of which are sufficiently well preserved to admit of identification. Those which are known are only a few thousands in number, and a mere fragment of the vast series of species which *have* existed in successive past ages of the earth. They are a few samples of the predecessors of the existing species, and some of them were the actual ancestors of those existing to-day. The larger number of them have left no direct issue, but represent side branches of the "tree of life" which have died out ages ago.

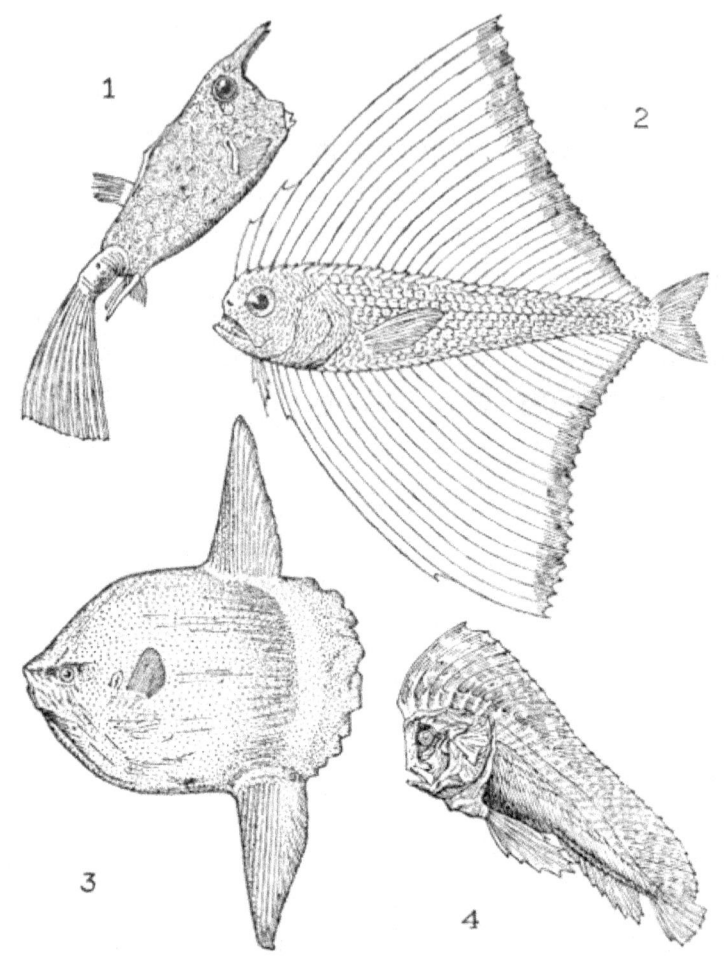

STRANGELY-SHAPED FISHES.—1. The Coffer-fish (Ostracion); 2. Pteraclis, an oceanic fish allied to the so-called Dolphins; 3. The Sun-fish (Orthagoriscus); 4. An Australian Blenny Patæcus.

CHAPTER XI

HYBRIDS

THE subject treated in this and the next chapter is one of the most interesting to mankind, and is surrounded by extraordinary prejudice, sentiment, and ignorance. It is one upon which really trustworthy information is to a very large extent absent—and difficult to obtain. I cannot profess to supply this deficiency, but I can put the matter before the reader.

It is a well-established fact that the various "kinds" of animals and of plants do not breed promiscuously with one another. The individuals of a "species" only breed with other individuals of that "species." They do not even, as a habit, breed with the individuals of an allied species. So nearly universal is this rule that it was for a long time held by naturalists to be an absolute definition of "a species," that it is a group of individuals capable of producing fertile young by breeding with one another and incapable of producing fertile young by mating with individuals of another such group, which were, therefore, held to constitute a distinct species. The practical importance of this definition was that it could, in a large number of instances among animals, and still more amongst plants, be made use of as a test and decided by experiment.

It is a curious fact that popular belief amongst country-folk and those who have opportunities of coming to a conclusion on so simple and direct a question has never accepted this law of the limitation of species in breeding as more than a general rule to which it has always been supposed that frequent exceptions occur. I mention this not in order to add that "there is always some basis of truth in these popular beliefs," but on the contrary to point out that popular beliefs on such matters are very frequently altogether erroneous, and though their origin can sometimes be explained, it is rare to find that they are due, in however small a degree, to true observation and inference. Where the subject under consideration has the obscurity and strong fascination for the natural man which all that relates to the processes of life, growth, and reproduction possess, we find that traditional fancies of the most unwarrantable kind are current, and hold their ground with tenacity even at the present day. Some 250 years ago, and earlier—in fact, before the commencement of that definite epoch of "the New Philosophy" marked by the foundation of the Royal Society of London—any queer-looking animal brought from remote lands, and any misshapen monstrosity born of cattle, sheep, dogs, or men, was "explained," and confidently regarded as a "hybrid," the result of a "cross" or irregular coupling of two distinct species of animals to which the "monster" presented some fanciful resemblance. Whole books were devoted to the description and picturing of such supposed examples of mis-begotten progeny.

The belief in the existence of such extraordinary hybrids is still common among so-called "well-educated" people. I have with difficulty avoided causing annoyance and offence to a friend, a celebrated painter, by refusing to admit that a deformed cat, of which he gave me an account, was a hybrid between a cat and a rabbit. A very eminent person whom I was conducting some years ago round the galleries of the Natural History Museum, declared, as we stood in front of the specimen of the Okapi of the Congo Forest, that it was clearly a hybrid between the giraffe and the zebra. He insisted that it was obvious that such was its explanation, and pointed to its striped haunches and legs, and its cloven hoofs and giraffe-like head. I failed to change his opinion.

It is the fact—ascertained by careful observation of natural occurrences and by experiment—that, in spite of the almost absolute law or general truth to the effect that the members of a species (whether of plant or animal) only produce fertile offspring by mating with members of that same species, yet there are rare instances known in which individuals of two distinct but allied species have mated and produced fertile offspring. The cases in which such unions have resulted in the production of offspring, but in which the offspring so produced prove to be infertile—that is, incapable of producing offspring in their turn—are much more numerous. An important distinction has also to be made between cases of either fertile or infertile hybrid-production which occur spontaneously in nature, and those in which man by separating the parent animals or plants from their natural conditions of life, or by bringing about impregnation (as in "pollinating" one flower with the pollen-dust of another) succeeds in obtaining a "cross" or "hybrid," whether fertile or infertile, not known to occur in "wild" (that is, not humanly controlled) nature. The rarest case would be that of the production of fertile hybrids in uncontrolled natural conditions. Such possibly occur in the case of some fishes in which the fertilization of the eggs takes place in water, the fertilizing microscopic sperms passing from the males like dust into the water and thus reaching the eggs laid by the females. Occasionally hybrids are thus produced between some common fresh-water fishes—species of the same genus—and between species of flat-fish, such as the turbot and the brill, though it is difficult to be sure that the rare hybrids so produced are fertile even if they attain to maturity. The same is true as to certain small flowering plants having distinct regions of natural distribution and occurrence. At the confines of the regions proper to two such allied species, insects passing from one to the other do sometimes effect a reciprocal fertilization of the two species, and a natural hybrid is the result. Here, again, it is difficult to follow the subsequent history of the hybrids, but it is believed that in some instances they are fertile, and that the hybrid race is only gradually merged by subsequent crossing into one or other of the parent species. Not a single instance is on record of the production of a "natural" hybrid (that is to say, one produced in natural conditions without man's interference), whether fertile or infertile, between two species of the larger animals (such as between horse and ass or zebra and ass, or between lion and tiger or any of the species of cats, or between species of bears) or birds (such as pheasants of various species, including the jungle cock, the wild original of our domestic fowl, or between various species of ducks, various species of geese, or between various species of the grouse-birds).

Nevertheless, in conditions brought about by man—that is to say, confinement in cages or paddocks, or at any rate removal from their native climate and home—all the groups of species just cited commonly and frequently produce hybrids *inter se*, that is, one or more species of the horse group thus inter-breed with one another, so will certain species of cats, certain species of bears, many species of pheasants, also of ducks, of geese, and of grouse. In nearly every case the hybrids so produced are infertile; they will not mate with a similar hybrid, and even when mated with one of the parent species rarely produce offspring, though they sometimes do so. The best cases of the production of fertile hybrids are between species of flowering plants brought to this country from widely separated regions. The surprising and instructive result has been obtained that a cross between two allied species (that is, of one and the same "genus") which will fail altogether or "come to nothing" as infertile hybrids—if the two species crossed are from the same or contiguous regions—yet will yield readily vigorous fertile hybrid offspring when the two species (always, of course, of one and the same genus) have their native homes in widely separate parts of the world—as, for instance, the Indian Himalaya range and the South American Andean range.

This has been found in crossing species of rhododendrons, of orchids, and of many other plants with which horticulturists occupy themselves for commercial purposes. It is in some ways the reverse of what one might expect. It would be reasonable to suppose that allied species from the same climate and geographical region would have more affinity and be more readily hybridized than species from widely remote and physically differing regions. But the reverse is the case, many thriving hybrid stocks which duly fertilize and set their seed are now in cultivation, having been produced by the union of parent species from "the opposite ends of the earth."

The consideration of this case throws some light on the significance of the non-occurrence of natural hybrids and of the very remarkable and curious fact that hybrids are so usually sterile. When we come to think of it, the natural preliminary assumption should be (as is that of unsophisticated humanity) that any animal or plant might, so far as possibilities go, breed with any other; and the questions to be answered are: (1) What advantage to a species is it not to be able to hybridize with other species, and (2) how—that is to say, by what structure or by what subtle chemical differences or other features in their make-up and habit—are they prevented from so hybridizing? Then we come on further to the question, Why should a hybrid, once produced, fail to bear healthy eggs or sperms according to its sex, although it grows up to full size and is to all appearances mature? And why should hybrids between parents of origin locally remote from one another not show this failure, but behave like ordinary healthy organisms?

In the full solution of these inquiries we should get very near to some of the most important secrets of the living body which have still to be searched out. But a reply to these questions which is probably in large measure true, and serves to help us in the further collection and examination of facts, is as follows: First, the production and maintenance of "species" of plants and of animals by survival of favourable variations in the struggle for existence (Darwin and Wallace's theory of the origin of species) requires the maintenance of the purity of the favourable stock which survives in the struggle. If it were continually liable to hybridization by other species it would never establish its own distinctive features. It would deteriorate by departing from those characteristics which have been "naturally selected" and have rendered it a successful "species." Thus the breeder, when he has selected a stock for propagation which approaches the standard at which he is aiming, keeps it apart, and does not allow it to be "crossed" by other stock. One of the qualities "naturally selected" in "the wild" is the power of resistance to fertilization by neighbouring species.

This power of resistance or immunity to fertilization by other species may be attained by several different methods. Amongst these are (1) a difference in the season of breeding or sexual ripening; (2) the production of secretions (whether by plant or by animal) which poison or paralyse the fertilizing sperms of allied and locally associated species, but are harmless to those of the secreting species; (3) the mechanical differences of size, etc., which prevent the fertilizing material of a strange species from gaining access to the egg-cells; (4) psychical activities (antipathies) in the case of animals or mere attraction and repulsion by odoriferous substances, which serve to repel a strange species, but are attractive to individuals of the same species; (5) finally, a chemical and physiological incompatibility between the sperms of one species and the germs of another (as distinct from the attraction or repulsion of the entire living individual), which, even when all other difficulties are absent or have been overcome, may be, and frequently is, present, so that the spermatozoon cannot penetrate the egg-cell even when resting upon it, but may be paralysed or repelled, and in any case is not guided and drawn into the aperture of the egg-covering, called the micropyle, or "little entry," so as to fuse with and fertilize the egg.

The operation of these hindrances to hybrid fertilization and breeding have been ascertained in several different instances. It is not always possible, and certainly not easy to ascertain, which is at work in any and every case. But we can well conceive that one or other of these agencies have been developed and accentuated by survival of the fittest, so as to protect a species against fertilization by a neighbouring species, and thus to enable it to maintain its own "bundle of characteristics" free from the swamping effects of "mixture" (that is, "hybridization") with another species. It is also thus intelligible that an allied species from a distant land against which our native species and its closer ancestry—struggling for purity of race—have had no occasion or opportunity to develop a repelling protection—will have no such difficulty in effecting the fertilization of the native species as have those adjacent species against whose intrusions the latter is specifically moulded and selected by long generations of severe natural selection.

The failure of hybrids generally to ripen their ova and sperm so as to reproduce themselves is a subject upon which, considering its enormous importance and significance, singularly little has been done in the way of investigation. Fifty years ago it was usually taught that the mule, between the horse and the ass, so largely produced under human superintendence for transport service, was unable to breed owing to some deformity in the reproductive passages. Even now no adequate study of the subject has been made, but it appears that whilst a female mule can be, and sometimes is, successfully mated to a horse or an ass, giving birth to a foal, the male mule does not produce fully-formed spermatozoa. What precisely is the nature of this failure, what the ultimate microscopic condition of the sperm cells in infertile male mules, or in any other infertile male hybrids, has not yet been properly worked out by modern cytological methods. It would be a matter of vast interest to determine what is the difference in the structure of the sperm-cells of a fertile and of an infertile male hybrid. At present, so far as I know, this has not been done.

So far what I have written applies to hybridization—the inter-breeding of distinct species. A similar but by no means identical subject is that of the inter-breeding of distinct races or varieties of one species, and the production of "mongrels." "Mongrels" are to races what "hybrids" are to species. To this branch of the subject belongs the study of the effects of intermarriage between distinct races of men.

CHAPTER XII

THE CROSS-BREEDING OF RACES

WE have seen that there is no simple rule as to the "mating" of individuals of a species with individuals of another closely allied but distinct species. Such mating very rarely comes about in natural conditions, but man by his interference sometimes succeeds in procuring "hybrids" between allied species. Hybrids between species belonging to groups so different as to be distinguished by zoologists as distinct "families" or "orders" are quite unknown under any circumstances. Such remoteness of natural character and structure as is indicated by the two great divisions of hoofed mammals—the even-toed (including sheep, cattle, deer, antelopes, giraffes, pigs and camels), and the odd-toed (including tapirs, rhinoceroses, horses, asses and zebras) is an absolute bar to inter-breeding. So, too, the carnivora (cats, dogs, bears and seals, and smaller kinds) are so remote in their nature from the rabbits, hares and rats—called "the rodents"—that no mating between members of the one and the other of these groups has ever been observed, either in nature or under artificial conditions.

Even when individuals of closely allied species mate with one another it is a very rare occurrence that the hybrids so produced ripen their ova and sperms so as to be capable of carrying on the hybrid race, though sometimes they do ripen them and breed. The great naturalist Alfred Wallace, in his most valuable and readable book called "Darwinism," expressed the opinion that the apparent failure of hybrid races to perpetuate themselves by breeding was to a large extent due to the small number of individuals used in experiments on this matter, and the in-

and-in breeding which was the consequence. One of the great generalizations established by Darwin is that in-and-in breeding is, as a rule, resisted in all animals and plants, and leads when it occurs to a dying-out of the inbred race by resulting feebleness and infertility. A large part of Darwin's work consisted in demonstrating the devices existing in the natural structure and qualities of plants and animals for securing cross-fertilization among individuals of the same species but of different stock. Both extremes seem to be barred in nature—namely, the inter-breeding of stocks so diverse in structure and quality as to be what we call "distinct species," and again the inter-breeding of individuals of the same immediate parentage or near cousinship. What seems to be favoured by the natural structure and qualities of the plant or the animal is that it shall only breed within a certain group—the species—and shall within that group avoid constant self-fertilization or fertilization by near cousins. Thus we find numerous cases in which, though the same flower has both pollen and ovules, and might fertilize itself, the visits of insects (specially made use of by mechanisms in the flower) carry the pollen of one flower to the ovules of another and to flowers on separate plants growing at a distance. It is necessary to note that there are, nevertheless, self-fertilizing flowers, and also self-fertilizing lower animals, the special conditions of which require and have received careful examination and consideration, upon which I cannot now enter.

In relation to the question of the possibility of establishing hybrids between various species experimentally, I must (before going on to the cognate question of "mongrels") tell of an interesting suggestion made to me by my friend Professor Alphonse Milne-Edwards not long before he died, and never published by him. He was director of the Jardin des Plantes in Paris, where there is a menagerie of living beasts as well as a botanic garden and great museum collections and laboratories. He held it to be probable, as many physiologists would agree, that the fertilization of the egg of one species by the sperm of another, even a remotely related one, is ultimately prevented by a chemical incompatibility—chemical in the sense that the highly complex molecular constitution of such bodies as the anti-toxins and serums with which physiologists are beginning to deal is "chemical"—and that all the other and secondary obstacles to fertilization can be overcome or evaded in the course of experiment. He proposed to inject one species by "serums" extracted from the other, in such a way as seemed most likely to bring the chemical state of their reproductive elements into harmony, that is to say, into a condition in which they should not be actively antagonistic but admit of fusion and union. He proposed, by the exchange of living or highly organized fluids (by means of injection or transfusion) between a male and female of separate species, to assimilate the chemical constitution of one to that of the other, and thus possibly so to affect their reproductive elements that the one could tolerate and fertilize the other. The suggestion is not unreasonable, but would require a long series of experiments in which the possibility of producing such "assimilation," even to a small extent and in respect of less complex processes than those ultimately aimed at, would have to be, first of all, established. My friend did not live to commence this investigation, but it is possible that some day we may see the obstacle to the union of ovum and sperm of species, which are to some extent allied, removed in this way by transfusion or injection of important fluids from the one into the other.

We must not lose sight of the fact, in the midst of these various and diverging observations about the fertilization of the ova of one species by sperms of another species, that there is such a thing as "parthenogenesis," or virgin-birth. In some of the insects and lower forms of animals the egg-cell habitually and regularly develops and gives rise to a new individual without being fertilized at all. And in other cases by special treatment, such as rubbing with a brush, or in the case of marine animals by addition of certain salts to the water in which the eggs are floating—or, again, in the case of the eggs of the common frog by gently scratching them with a needle—the eggs which usually and regularly require to be penetrated by and fused with a spermatozoon or sperm-filament before they will develop, proceed to develop into complete new individuals without the action upon them of any spermatozoon. In such marine animals as the sea-urchins or sea-eggs it has been found that the eggs deposited in pure sea-water, though they would die and decompose if left there alone, can be made to develop and proceed on their growth by the addition to the sea-water of the sperm filaments of a star-fish (the feather star or comatula). The spermatozoa or sperm-filaments do not, however, in this case fuse with the egg-cells. They mechanically pierce the egg-coat, but contribute no substance to the embryo into which the egg develops. They have merely served, like the scratch of a needle on the frog's egg and the brushing of insects' eggs, to start the egg on its growth, to "stimulate" it and set changes going. It appears thus that the fertilizing sperm-filaments of organisms generally have two separate and very important influences upon the egg-cells with which they fuse. The one is to stimulate the egg and start the changes of embryonic growth; the other is to contribute some living material from the male parent to the new individual arising from the growth and shaping of the egg-cell. The first influence can be exercised without the second, as is seen in the case of the eggs of some sea-urchins stimulated to growth by the spermatozoa of some star-fishes. It happens that these marine animals are convenient for study and experiment because their eggs are small and transparent and that they and the spermatozoa are freely passed into the sea-water at the breeding season, in which the fertilization of the eggs takes place.

When these facts are considered we have to admit that in the mating of two species which will not regularly and naturally breed together, there may be a limited action of the spermatic element which may stimulate the egg to development without contributing by fusion in the regular way to the actual substance of the young so produced, or only contributing an amount insufficient to produce a full and normal development of the hybrid young. Such cases not improbably sometimes occur in higher animals, though they have not been, as yet, shown to exist except in the experiments with sea-urchins' eggs and feather-star's sperm.

In all animals and plants, but especially in domesticated and cultivated stocks or strains, varieties arise which, by natural or artificial separation, breed apart, and give rise to what are called "races." Such races in natural conditions may become species. Species are races or groups of individuals, which, by long estrangement (not necessarily local isolation) from the parent stock and by adaptation to special conditions of life, have become more or less "stable"—that is, permanent and unchanging in the conditions to which they have become adapted. They acquire by one device or another the habit of not breeding with the stock from which they originally diverged—a repugnance which may be overcome by human contrivance or by natural accident, but is, nevertheless, an effective and real quality. Distinct forms, which have not arrived at the stability and separation characteristic of species, are spoken of as "races," or "varieties." It is very generally the case that the "races" of one species can inter-breed freely with one another, and with the original stock, when it still exists. Comparatively little is known as to the behaviour of wild or naturally-produced "races." Practically all our views on the subject of "races" and their inter-breeding are derived from our observation of the immense number and range of "races" and "breeds" produced by man—as farmer, fancier, and horticulturist. It has been generally received as a rule, that the various races produced in the farm or garden by breeding from a species, will inter-breed freely, and produce offspring which are fertile. A special and important series of races, in which human purpose and voluntary selection necessarily have a leading part, are the races of man.

The offspring of parents of two different races is called a mongrel, whilst the term "hybrid" has been of late limited, for the sake of convenience, to the offspring of parents of two different species. Mongrels, it has been generally held, are fertile—often more fertile than pure-bred individuals whose parents are both of the same race, whilst "hybrids" are contrasted with them, in being infertile. We have seen that infertility is not an absolute rule in the case of hybrids, and it appears that there is also a source of error in the observations which lead to the notion that "mongrels" are always fertile. The fact is that observations on this matter have nearly always been made with domesticated animals and plants which are, of course, selected and bred by man on account of their fertility, and thus are exceptionally characterized by fertility, which is transmitted in an exceptional degree to the races or varieties which are experimentally inter-bred, and, consequently, may be expected to produce fertile mongrels. Alfred Russel Wallace insisted upon this fact, and pointed out that in a few cases colour varieties of a given species of plant have been found to be incapable of inter-breeding, or only produce very few "mongrels." This has been established in the case of two dissimilarly-coloured varieties of mullein. Also the red and the blue pimpernel (the poor man's weather-glass, Anagallis), which are classed by botanists as two varieties of one species, have been found after repeated trials to be definitely incapable of inter-breeding. Wallace insists in regard to crossing, that some degree of difference favours fertility, but a little more tends to infertility. We must remember that the fertility of both plants and animals is very easily upset. Changed conditions of life—such as domestication—may lead (we do not know why) to complete or nearly complete infertility; and, again, "change of air," or of locality, has an extraordinary and not-as-yet-explained effect on fertility.

"Oh, the little more and how much it is!
And the little less, and what worlds away!"

Infertile horses sent from their native home to a different climate (as, for instance, from Scotland to Newmarket) become fertile. A judicious crossing of varieties or races threatened with infertility will often lead to increased vigour and fertility in the new generation, just as change of locality will produce such a result. Physiological processes which are not obvious and cannot be exactly estimated or measured are then, we must conclude, largely connected with the question of sterility and fertility. Mr. Darwin has collected facts which go far to prove that colour (as in the case of the black pigs of Virginia, which I cited in Chapter X.), instead of being a trifling and unimportant character, as was supposed by the older naturalists, is really one of great significance, often correlated with important constitutional differences. It is pointed out by Alfred Wallace that in all the recorded cases in which a decided infertility occurs between varieties (or races) of the same species of plants (such as those just cited), those varieties are distinguished by a difference of colour. He gives reasons for thinking that the correlation of colour with infertility which has been detected in several cases in plants may also extend to animals in a state of nature. The constant preference of animals—even mere varieties of dog, sheep, horses, and pigeons—for their like, has been well established by observation. Colour is one of the readiest appeals to the eye in guiding animals in such selection and association, and is connected with deep-seated constitutional qualities. "Birds of a feather flock together" is a popular statement confirmed by the careful observation of naturalists. Thus we arrive at some indication of features which may determine the inter-breeding, or the abstention from inter-breeding, of diverse races sprung from one original stock. The "colour bar" is not merely the invention of human prejudice, but already exists in wild plants and animals.

We now come to the questions, the assertions, the beliefs, and the acts concerning the inter-breeding of human races, to the consideration of which I have been preparing the way. The dog-fancier has generally a great contempt for "mongrels." Breeders generally dislike accidental crosses, because they interfere with the purpose which the breeder has in view of producing animals or plants of a quality, form, and character which he has determined on before-hand. This interference with his purpose seems to be the explanation of beliefs and statements, to the prejudice of "mongrels." Really, as is well known to great breeders and horticulturists, a determined and selective crossing of breeds is the very foundation of the breeder's art, and there is no reason to suppose that a "mongrel" is necessarily, or even probably, inferior in vigour or in qualities which are advantageous in the struggle for life in "natural"—that is to say, "larger"—

conditions of an animal's or plant's life; not those limited conditions for which the breeder intends his products. Indeed, the very opposite is the case. In nature, as Mr. Darwin showed, there are innumerable contrivances to ensure the cross-breeding of allied but distinct strains. Dog-owners who are not exclusively bent upon possessing a dog which shows in a perfect way the "points" of a breed favoured by the fashion of the moment, or fitting it for some special employment, know very well that a "mongrel" may often exhibit finer qualities of intelligence, or endurance, than those exhibited by a dog of pure-bred "race." And the very "races" which are spoken of to-day as "pure-bred," or "thoroughbred," have (as is well known) been produced as "mongrels"—that is to say, by crossing or mating individuals of previously-existing distinct and pure breeds. The history of many such "mongrel breeds," now spoken of as "thoroughbred," is well known. The English racehorse was gradually produced by the "mongrelizing," or cross-breeding, of several breeds or races—the English warhorse, the Arab, the Barb. A very fine mongrel stock having at last been obtained, it was found, or, at any rate, was considered to be demonstrated, that no further improvement (for the purposes aimed at, namely, flat-racing) could be effected by introducing the blood of other stock. The offspring of the "mongrels" Herod, Matchem, and Eclipse accordingly became established as "the" English racehorse, and thenceforward was mated only within its own race or stock, and was kept pure or "thoroughbred." Another well-known mongrel breed which is now kept pure, or nearly so, is that of the St. Bernard's dog, a blend of Newfoundland, Bloodhound, and English Mastiff.

Often the word "mongrel" is limited in its use to signify an undesired or undesirable result of the cross-breeding of individuals of established races. But this is not quite fair to mongrels in general, since, as we have seen, the name really refers only to the fact they are crosses between two breeds. When they happen to suit some artificial and arbitrary requirement they are favoured, and made the starting-point of a new breed, and kept pure in their own line; but when they do not fit some capricious demand of the breeder they are sneered at and condemned, although they may be fine and capable animals. No doubt some mongrels between races differing greatly from one another, or having some peculiar mixture of incompatible qualities the exact nature of which we have not ascertained, are wanting in vigour, and cannot be readily established as a new breed. In nature the success of the mongrel depends on whether or not its mixture of qualities makes it fitter than others to the actual conditions of its life, and able to survive in the competition for food and place. In man's breeding operations with varieties of domesticated animals and "cultivated" plants, the survival of the mongrel depends upon its fitting some arbitrary standard applied by man, who destroys those which do not suit his fancy, and selects for survival and continued breeding those which do.

What is called "miscegenation," or the inter-breeding of human races, must be looked at from both these points of view. We require to know how far, if at all, the mixed or mongrel offspring of a human race A with a human race B is really inferior to either of the original stocks A and B, judged by general capacity and life-preserving qualities in the varied conditions of the great area of the habitable globe. And how far an arbitrary or fanciful standard is set up by human races, similar to that set up by the "fancier" or cultivator of breeds of domestic animals. The matter is complicated by the fact that what we loosely speak of as "races" of man are of very various degrees of consanguinity or nearness to one another in blood, that is, in stock or in ultimate ancestry. It is also complicated by the fact that we cannot place any reliance upon the antipathies or preferences shown by the general sentiment of a race in this (or other matters) as necessarily indicating what is beneficial for humanity in general or for the immediate future of any section of it. Nor have we any assurance that what is called "sexual selection"—the preference or taste in the matter of choosing a mate—is among human beings necessarily anything of greater importance—so far as the prosperity of a race or of humanity in general is concerned—than a mere caprice or a meaningless persistence of the human mind in favouring a choice which is habitual and traditional. I have referred to this point again in the last paragraph of this chapter.

In regard to marriage between individuals of different European nationalities, a certain amount of unwillingness exists on the part of both men and women which cannot be ascribed to any deep-seated inborn antipathy, but is due to a mistrust of the unknown "foreigner," which very readily disappears on acquaintance, or may arise from dislike of the laws and customs of a foreign people. English, French, Dutch, Scandinavians, Germans, Russians, Greeks, Italians and Spaniards have no deep-rooted prejudices on the subject, and readily intermarry when circumstances bring them into association. Though the Jews by their present traditional practice are opposed to marriage with those not of their faith, there is no effective aversion of a racial kind to such unions, and in early times they have been very frequent. During the "captivity" in Babylon and again after the "dispersal" by the Romans, the original Jewish race was practically swamped by mixture with cognate Oriental races who adopted the Jewish faith. So far from there being inborn prejudice against intermarriage of the peoples above cited, it is very generally admitted that such "miscegenation" leads frequently to the foundation of families of fine quality. The blend is successful, as may be seen in the number of prominent Englishmen who have Huguenot, German, Dutch, or Jewish blood in their veins.

But when we come to the intermarriage of members of the white race of Europe with members of either the negroid (black) race or of the yellow and red mongoloid race, a much greater and more deeply-rooted aversion is found, and this is extended even to members of the Caucasian race who, possibly by prehistoric mixture with negro-like races, are very dark-skinned, as is the case with the Aryan population in India and Polynesia. It is a very difficult matter; in fact, it seems to me not possible in our present knowledge of the facts, to decide whether there is a natural inborn or congenital disinclination to the marriage of the white race, especially of the Anglo-Saxon branch of it, with "coloured" people, or whether the whole attitude (as I am inclined to think) is one of "pride of race," an attitude which can be defended on the highest grounds, though it may lead to erroneous beliefs as to the immediate evil results of such unions, and to an unreasonable and cruel treatment both of the individuals so intermarrying and of their offspring. There is little or no evidence of objection to mixed unions on the part of the coloured people with whites, no evidence of physical dislike to the white man or white woman, but, on the contrary, ready acquiescence.

A curious aversion to marriages with whites on the part both of North American Indians and of negroes is, however, recorded from time to time in the official reports of the United States Government.

Two beliefs about such unions are more or less prevalent among white men in the regions where they not infrequently occur. Neither of these beliefs is supported by anything like conclusive evidence. The one is that such unions lead to the production of relatively infertile offspring; the mixed breed or stock is said to die out after a few (some seven or eight) generations. It is, however, the fact that the circumstances under which this occurs suggest that it is not due to a natural and necessary infertility. The other assertion is that the offspring of parents—one of white race and the other of black, yellow or brown—tend by some strange fatality to inherit the bad qualities of both races and the good qualities of neither. This is a case to which must be applied the saying, "Give a dog a bad name and hang him." The white man in North America, in India, and in New Zealand desires the increase and prosperity of his own race. Like the fancier set on the production of certain breeds of domesticated animals, he has no toleration for a "mongrel." In so far as it is true that miscegenation (marriage of white and coloured race) produces a stock which rapidly dies out—this is due to the adverse conditions, the opposition and hostility to which the mixed race is exposed by the attitude of the dominant white race. To the same cause is due the development of ignoble and possibly dangerous characteristics in the unfortunate offspring of these marriages more frequently than in those who find their natural place and healthy up-bringing either in the white or the coloured sections of the community. The "half-breed" is in some countries inexorably rejected by the race of his or her white parent and forced to take up an equivocal association with the coloured race.

That some, at any rate, of the evils attributed to "miscegenation" are due to the baneful influence of "pride of race" is evident from the fact that the Portuguese (with the exception of a small aristocratic class) have not since the early days of the fourteenth century, perhaps in consequence of established association with the Moorish and other North African races, shown that pride of race and aversion to mixture with dark-skinned races which is so strong a feature in the Anglo-Saxons, their successors and rivals as colonists. The long-standing admixture of black blood in the Portuguese population before the colonization of South America, has led to a toleration on the part of the Portuguese colonists of "miscegenation," both with Indians and the liberated descendants of imported negro slaves. The consequence is that in Brazil there is no condemnation of black blood; children of mixed parentage and of coloured race attend the same schools as those of European blood, and freely associate with them. There is no notion that that portion of the population which is of mixed negro, Indian, and white blood is less vigorous or fertile than the unmixed, nor that vice and feebleness are the characteristics of the former, whilst virtue and capacity belong to the latter.

The determined hostility of the Anglo-Saxon race in North America and in British India to "miscegenation" is in the case of the United States to be explained by the peculiar relation of a large slave population in the Southern States to a pure white slave-owning race: whilst in India we have a handful of white men temporarily stationed as rulers of millions of "natives," but never accepting India as their home. The attitude of the Anglo-Saxon race to the North American Indians, and also to the Maoris of New Zealand, has never been so extreme in the matter of miscegenation as it has been to negroid people and to the very different though dark-skinned people of the East. In support of that opinion may be cited the fact that some of "the first families of Virginia" are proud of their descent from Pocahontes, the Algonkian "Princess" who married the Englishman Rolfe. In New Zealand there are many families of mixed Anglo-Saxon and Maori blood. Though they are not ostracized, as are the half-breeds of negro blood in the United States, there is a firm tendency to relegate the half-breeds in New Zealand to the Maori section of the population, which it must be remembered includes some of the richest and most prosperous landowners in the colony.

It may be questioned whether there is in this matter a greater "pride of race" among Anglo-Saxons than among other Northern European peoples. Neither the French nor the Germans have established great colonies like the English, nor undertaken the administration of a huge Eastern Empire, and have, therefore, not shown what attitude they would adopt under such circumstances. The tolerance and easy-going humanitarianism of the French in relation to "miscegenation" in their dependencies in past times has never had the significance or practical importance which it would have possessed in the English Colonies and in the great Indian Empire.

There is, on account of the sporadic and exceptional occurrence of modern instances, no information of any value as to the results of mixture of other races of man. In early times and among more primitive or less civilized peoples there appears to have been, when immigration or conquest gave the opportunity, no obstacle to a free intermixture of an incoming race with the natives of an invaded territory. The "pride of race" has, nevertheless, throughout historic time been a frequent factor in the adjustment of populations of diverse races, and though "colour" has been a frequent "test" or symbol of the superior and exclusive race, it has not been the only characteristic exalted to such importance. Such "pride of race" has frequently excluded the members of a closely allied but conquered racial group from intermarriage with the conquerors, and has only disappeared after centuries of persistence. The term "blue blood" is interesting in this connection. It is the "saing d'azure" of the Gothic invaders, the conquerors of the Iberian and Moorish people of Spain. It refers not to any "blueness" of the blood itself, such as distinguishes veinous from arterial blood, but to the blue colour of the veins as seen through the colourless skin of a northern race (the Goths), as compared with the invisibility of the veins when the skin is rendered more or less opaque by a brown pigment, as in the Moors and the swarthy Iberians.

Among the people of Western Europe marriage has assumed more and more a character which is almost unknown in the rest of the world. Whatever the future may be in regard to this matter, there is no doubt possible that the place given to women in Western Europe by the ideals of chivalry and the practice of the northern race (which has so largely displaced the traditions of the Roman Empire) has established a relation of the sexes in which marriage and consequent parentage have ceased to be regarded as a mere regularization of animal desire and appetite. The accepted, but not always consciously recognized, view of marriage in Western Europe is that the union so sanctioned and the families thereby produced should be the result not of the mere physical necessity of irresponsible victims of an impulse common to all animals, but the outcome of the deliberate choice of man and woman attracted to one another by sympathy, understanding and reciprocal admiration, based upon knowledge of character, mental gifts and aspirations, as well as upon bodily charm. A rarely-expressed but none the less deeply-seated conviction exists that from such unions children of the finest nature, nurtured in circumstances most likely to make them worthy members of the community, will be born and reared. It is this conviction which leads to, or at any rate endorses, the exclusiveness which is described as "pride of race." The Anglo-Saxon man and equally the

Anglo-Saxon woman (as well as the allied races of neighbouring nationalities) recognize a responsibility, a race duty, resulting from accumulated tradition, the heirloom of long ages of family life, which causes the man to be ashamed of, and the woman to shrink with instinctive horror from, union with an individual of a remote race with whom there can be no real sympathy, no intimate understanding. That seems to me to be the explanation and the justification of the "colour bar."

In relation to the probable effectiveness of sexual selection among uncivilized peoples in favouring and maintaining a particular type or form of features, hair, etc., characteristic of the race, independently of the life-preserving value of such qualities, I may mention, before quitting this difficult but strangely fascinating subject, a fact observed by a traveller in Africa, and related to me by him. Other similar facts are on record. Among the negroes employed as "porters" by my friend, some thirty in number, was one who had a narrow aquiline nose and thin lips. He was as black and as woolly-haired as any of them, but would if of fair complexion have been regarded by Europeans as a very handsome, fine-featured man. Such cases are not uncommon in parts of Africa, where probably an unrecognized mixture with Arab or Hamite blood has occurred. My friend expected this man to be a favourite, on account of what to him appeared to be "good looks," with the girls of the villages at which he camped during a three months' journey. At every such village, as they journeyed on, the travellers received with joy and good nature. The negro porters were fêted and made much of by the young women. But one alone was unpopular and regarded with ridicule and dislike. This was the handsome negro with the fine, well-modelled nose and beautiful European lips. The black beauties turned their backs on him, in spite of his amiable character and kindly overtures.

They invariably and by open confession preferred the men with the thickest lips, the broadest noses, and the most thoroughly (as we should say) degraded prognathous appearance and disgusting expression. Hence no doubt the young negresses were likely to perpetuate in their offspring the features which are characteristic of their race, and hence it is probable that mere capricious sexual selection of individuals most completely conforming to a preferred type—irrespective of the value of the features preferred—may have great effect in both the selection and the maintenance of the peculiarities of the type. Dark skin may thus have been selected, until it became actually black; a slight curling of the hair, until it became woolly; thickish lips and broadish nose, until they became excessive in thickness and breadth.

CHAPTER XIII

WHEEL ANIMALCULES

TWO hundred years ago the Dutch naturalist Leuwenhoek, who made many discoveries with the highly magnifying lenses which he himself ground and mounted, wrote to the Royal Society of London that he had "discovered several animalcula that protrude two wheels out of the forepart of their body as they swim, or go on the sides of the glass jar in which they are living." He says that "the two wheels are thick set with teeth as the wheel of a watch," and he sent to the society for publication drawings of these wonderful little creatures. This was the first account of the Wheel Animalcules. Since then they have been studied by many microscopists, especially by Ehrenberg, who figured many in his great book on animalcules in 1838. Fourteen years later the delightful English naturalist, P. H. Gosse, who studied and illustrated the "sea-anemones" so ably—and, by his example and charming descriptions, made the keeping of these beautiful things in marine aquaria a favourite occupation among people of leisure, blessed with a "curiosity concerning the things of nature"—published some microscopical studies on Wheel Animalcules, and continued throughout his life to make them a special subject of his investigation.

The microscope was greatly improved—in fact, reached its present state of perfection—during Mr. Gosse's lifetime, and a wonderful amount was added to our knowledge not only as to the various kinds of wheel animalcules (which now number not less than 900 species), but also with regard to the minutest details of their structure, their growth from the egg, and their habits. Another English lover of these minute creatures, Dr. C. T.

Hudson, of Clifton (Bristol), began his observations a few years later, and also discovered many wonderful kinds. It was my good fortune to bring these two devotees of the Rotifera, or Wheel Animalcules, together, and to induce them to write a conjoint work on their favourites—after, as they say in their preface, they had each continued their studies almost daily for thirty years, and had made innumerable drawings from living specimens, which are reproduced in the many hundred (mostly coloured) figures engraved in the thirty-four quarto plates of their monumental book. This was published in 1889, a year after Mr. Gosse's death at the age of 78. My friend, Mr. Edmund Gosse, the distinguished man of letters, is the son of the naturalist; the microscope, the aquarium, and the rock-pools of the seashore were the familiar delights of his boyhood, as of mine.

In Fig. 34 I have sketched the common Rotifer or wheel animalcule. It is about the one-fortieth of an inch long. The two specimens drawn in Figs. 34, A and B, are seen to be clinging by the forked tail-end of the body to a piece of weed (drawn in dotted lines). The body is stretched in these specimens to its full length. It can be shortened by a "telescoping" or pulling in of either end, so as to make the animal a mere oval particle. The four narrower joints or segments at the tail-end can be pulled in like the segments of a telescope, whilst the two wheels and adjacent parts can be drawn down into the body as shown in Fig. 34, C, where the two wheels (W) are seen showing through the skin by transparency.

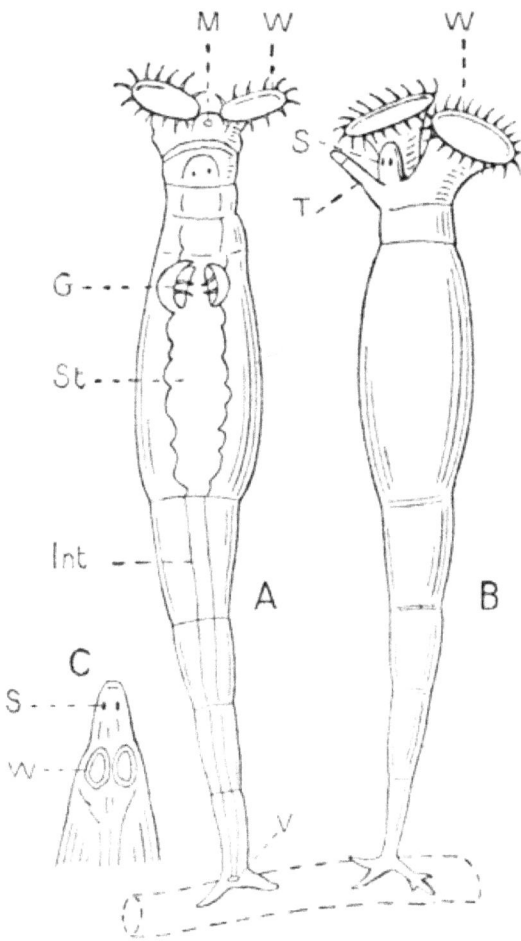

FIG. 34.—Diagram of *Rotifer vulgaris*—the common wheel
animalcule—one hundred and twenty times as long as the
creature itself. *A*, front view. *B*, side view. *C*, head showing eyes
S, and retracted wheel apparatus *W*. The letters in *A* and *B* have
the following signification: *M*, mouth. *W*, wheel or ciliated disc.
S, eye spots on head. *T*, spur or tentacle. *G*, gizzard. *St*, stomach.
Int, intestines. *V*, vent: aperture of intestine.

The common rotifer can walk like a looping caterpillar
or a leech—fixing itself by its tail, then stretching out the
head and fixing that, whilst letting go the tail and bringing
it up by "telescoping" it, near to the head region. The tail is
forked, and in the side view (Fig. 34, B) it is seen to have a
soft branched process, which helps it to cling. The letter V

in Fig. 34, A, points to the vent or opening of the gut at the fork of the tail. The mouth, marked M, is seen between the two "wheels." The two "wheels" are really two discs, the edges of which are beset by coarse "cilia," or vibrating hairs of protoplasm. [5] These cilia "lash" and straighten again one after the other, so that the optical illusion is produced of the toothed edge of the disc being in movement like a wheel. They may be "focused" with the microscope so that the groups or "bunches" of them look like stiff, motionless "teeth," although they are really, all the time, lashing and beating in regular rhythm. When the animal is fixed by its tail, the lashing of the cilia on the wheels causes currents in the water which set with great strength to the mouth and bring floating food particles to it. It is thus that the Rotifer feeds. When the tail is not grasping a support, the movement of the cilia on the wheels causes the animal to swim forward through the water, so that it has two modes of locomotion—the leech-like crawling method and the free swimming method.

The various internal organs of a Rotifer are readily seen through its transparent skin (Fig. 34, A). It has a nervous system, many bands of contractile muscles and a pair of little tubular kidneys or nephridia, besides reproductive germs (the eggs). I have here sketched only the digestive canal. The mouth leads through a gullet to a very curious organ called the "gizzard," marked G. All the wheel animalcules have this gizzard, but its teeth, shown as two oval bodies in the drawing, differ a great deal in shape and complexity in the different kinds. Whilst the Rotifer is feeding by bringing currents of water to its mouth, the two halves of the gizzard are kept in rapid movement by muscles, causing them to rub against one another and to grind up the food particles which reach them through the gullet. The gizzard (G) is followed by the digestive stomach (St), and that by the intestine (Int), which opens at the vent (V). The side (or three-quarter profile) view of a similar specimen (Fig. 34, B) shows only the surface of the little animal, and is intended to show especially the snout-like head-lobe (S), with its two eye-spots, which are red in colour. Standing out backwards from this is a finger-like process (T), which is called the spur, or tentacle. It has hairs at its tip, and is a sensory organ.

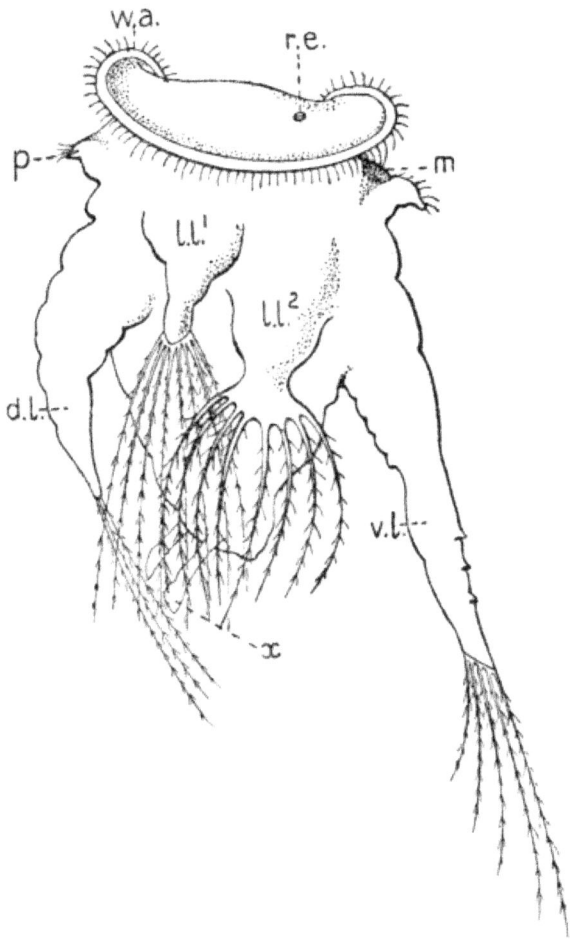

FIG. 35.—The Rotifer *Pedalion mirum*—seen from the right side, magnified 180 diameters. *w.a.*, wheel apparatus or "ciliated" margin of the cephalic disc. *r.e.*, right side eye-spot. *m.*, mouth. *p.*, tactile process. *d.l.*, median dorsal limb (as it is seen in profile, only three of the fringed hairs at its extremity are seen). *v.l.*, the great ventral limb (only five of its fan of eight fringed

hairs are seen). $l.l.^1$, dorso-lateral, and $l.l.^2$, ventro-lateral limbs of the right side: they show the complete fans of eight fringed hairs. $x.$, the pair of posterior processes tipped with vibratile cilia, better seen in Fig. 36.

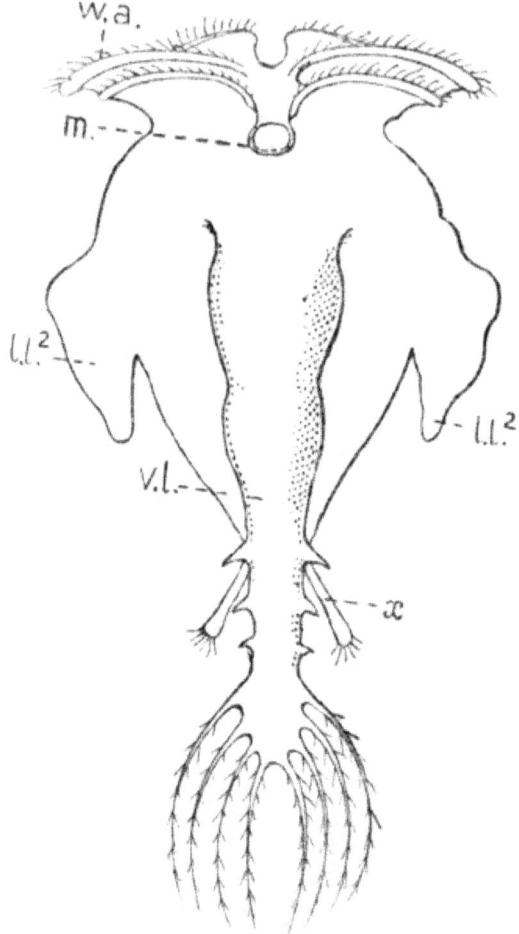

FIG. 36.—The Rotifer *Pedalion mirum*—seen from the ventral surface. Letters as in Fig. 35. The complete fan of eight fringed hairs terminating the great ventral limb are seen, and the three spine-like processes on each side of it. The fringed hairs of the two ventro-lateral limbs, $l.l.^2$, are omitted; they are fully shown in Fig. 35, and are the same in number and disposition as those forming the "fan" of the great ventral limb. Compare these hairs

with those of the "Nauplius" Crustacean larva drawn as a tail-piece to Chapter XIII.

In some wheel animalcules there is a pair of these spurs, and the very remarkable wheel animalcule drawn in Figs. 35 and 36 has six large processes which, though much bigger, appear to be of the same nature. Of these four are seen in Fig. 35, namely, *d.l.*, the dorsal limb, *v.l.*, the great ventral limb, and *l.l.*^{1} and *l.l.*^{2}, the two lateral limbs of the right side, all of them carrying fan-like groups of fringed hairs. They are moved by very powerful muscles, and strike the water with energetic strokes, so as to cause the little owner to dart through it. This jumping or darting wheel animalcule is called "Pedalion," and was discovered and described by Dr. Hudson. It is so astonishing and wonderful a little beast, that when Dr. Hudson sent me some alive in a tube by post in 1872, soon after he had discovered it, I could not believe my eyes, and thought I must be dreaming. It is very like the young form of Crustaceans known as a "Nauplius" (see tail-piece to the present chapter) in having (what no other wheel animalcule has) great hollow paired limbs moved by *striated* muscular fibre, carrying fringed hairs only known before in Crustaceans (crabs, shrimps and water fleas), and striking the water violently just as do those of the Nauplius. And yet all the while it has on its head a pair of large ciliated wheels which serve it just as do those of the common Rotifer. No Crustacean, young or old, has this "wheel-apparatus" nor any vibratile "cilia" on the surface of its body. Pedalion possesses an astounding "blend" of characters. Fig. 35 shows, besides the "paddles" or "legs" (of which two on the other side of the animal are not seen), the broad and large wheel-apparatus W (within which the right eye-spot *r.e.* is seen), and a little lobe (*p*) called the "chin" lying just below the mouth (*m*). The big leg (*v.l.*) and the pair on each side (*l.l.*^{1} and *l.l.*^{2}), of which that on the right side only is seen, end in beautiful fringed hairs, which are only seen elsewhere in the Crustacea (water-fleas and others). Those on the lateral limbs and the great ventral limb (Fig. 36) are set in two groups of four on each side of the free end of the limb, whilst those on the dorsal leg (*d.l.*) are apparently not so numerous. I have corrected the drawings, Figs. 35 and 36, by reference to actual specimens kindly given to me by Mr. Rousselet.

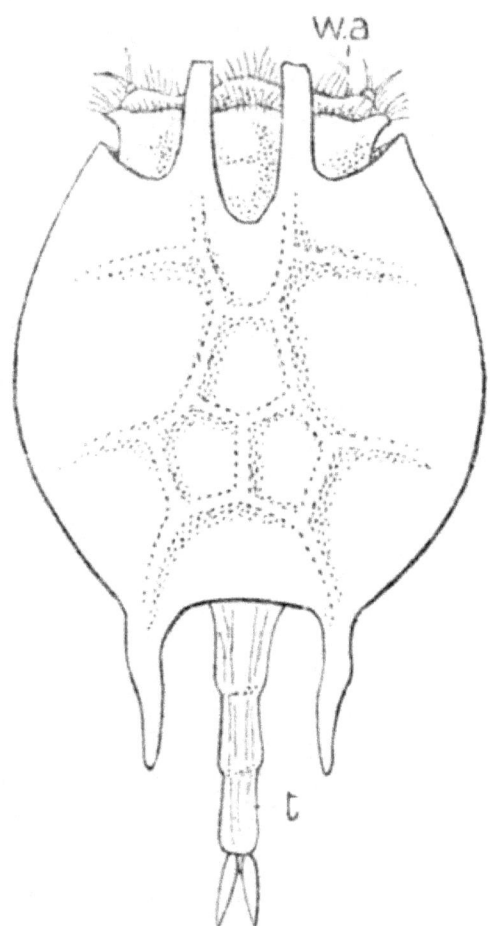

FIG. 37.—The Rotifer *Noteus quadricornis*—to show its curious four-horned carapace—from which the wheel apparatus, *wa,* emerges in front, and the tail, *t*, behind; somewhat as the head and tail of a tortoise emerge from its protective "box" or

carapace. The ridges on the horney covering of the Rotifer recall the horney plates of the tortoises and turtles.

The 500 different species of Wheel Animalcules or Rotifera differ from one another in the exact shape of the wheel-apparatus, in the jointing of the body and its general shape, and in the development, in some, of a hard skin or shell like a turtle's or tortoise's shell (Fig. 37) over that broadest region of the body in which in our Fig. 34, A, the stomach marked "St" is placed. They differ also in the shape of the gizzard's teeth, in the presence of paddles or legs (in Pedalion alone), and in the presence in some of longer or shorter projecting movable rods or bristles in pairs or in bunches. Many build for themselves tubular habitations of jelly or of hard cemented particles. They are all minute (from the $\frac{1}{12}$ to the $\frac{1}{500}$ in. in length). They are divided into five principal groups, which are (1) the crawlers, like the common Rotifer (Fig. 34), which can crawl like a leech and also swim freely by aid of their wheel-apparatus; (2) the naked free swimmers, which do not crawl, but move only by swimming; (3) the turtle-shelled free swimmers (Fig. 37) like the last, but provided with strong, often faceted, angular, and spike-bearing shells or "bucklers," from which head and wheel-apparatus project in front and narrow tail behind; (4) the rooted or fixed forms (Figs. 37 *bis*); these never swim when full grown, but each forms and inhabits a protective tube or case; (5) the skipping or darting forms. Of these there is only the Pedalion mirum (Figs. 35 and 36), which is quite unlike all the other wheel animalcules in having limbs like those of the minute water-fleas (Nauplius, Cyclops) which strike the water and are fringed with feather-like hairs.

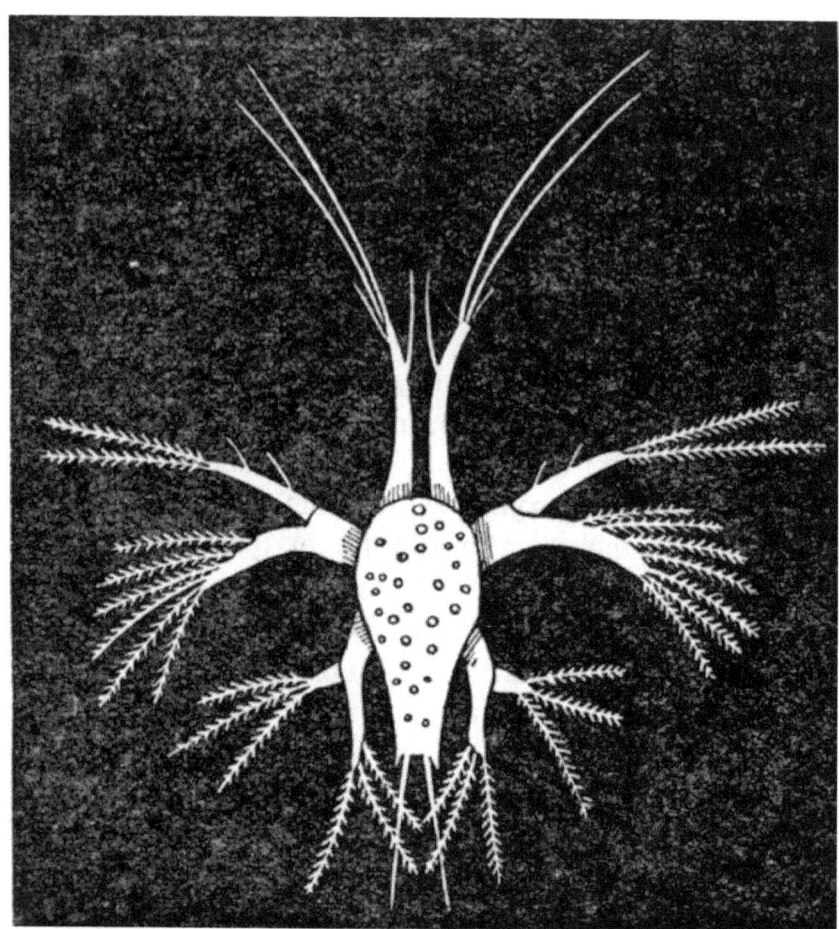

The larval or young form of Crustacea known as "the Nauplius." This is the "Nauplius" of a kind of Prawn. The three pairs of branched limbs are well seen. Much magnified.

FOOTNOTE:

[5] For some account of "cilia," see "Science from an Easy Chair," Figs. 29, 33, 40 and the accompanying text.

CHAPTER XIV

MORE ABOUT WHEEL ANIMALCULES

MICROSCOPIC as the wheel animalcules are they yet have been watched and examined by their admirers to as great a point of intimacy as that reached by the devotees of insects or of birds. A remarkable fact about them is that in about 130 different species (out of the 500 known) it has been found that the males are diminutive creatures, about one-tenth the size of the females, and devoid of digestive canal; in fact, little more than minute swimming sacs full of spermatozoa. In one group, that of the crawling Rotifers, to which the common wheel animalcule, figured in the last chapter, belongs, no male at all has ever been discovered. They are all females. They are precisely those wheel animalcules which are known to microscopists for their power of surviving (like the little water-bears or tardigrades and some other minute animalcules) the desiccation, or "drying-up" of the water in which they were living, swimming, and crawling (see Chapters XV. and XVI.). And it is quite probable that this power of resistance to the adverse conditions of changing seasons has, in the crawling Rotifers, taken the place of the production of eggs fertilized by a male. For, as in the case of the crustacean water-fleas (and of the terrestrial plant-lice, or aphides and gall-flies), it is found that the female Rotifers or wheel animalcules, which hatch from fertilized eggs, are themselves "parthenogenetic," and lay eggs which develop without fertilization by males—that is to say, are "impaternate." In the case of the water-fleas these are called "summer eggs," and after one or more generations of such fatherless females a proportion of males are produced which fertilize the females hatched at the same period. The eggs so fertilized acquire a thick shell and are called "winter eggs." They remain dormant for some months and resist the injurious influences of winter cold, or, it may be, of drying up and conversion of the pond-mud into dust, but hatch out when warmer and wetter conditions return.

This, however, is just what the adult crawling kind of Rotifer can do in the full-grown state by drawing up her body into the shape of a ball and exuding a jelly-like or horny coat. So that she has no need of "winter eggs," and the whole process of forming them and of males to impregnate them has "dropped out" of the life-history of this special kind of resistant Rotifers. The minute insignificant males and the eventual disappearance of males altogether in some races is a subject which may well occupy the attention of our human "suffragettes." That the males are minute creatures, less than the thousandth part of the size of the females, is a fact also ascertained in the case of some curious marine worms (called Bonellia and Hamingia). The only other instance of such degradation of the male sex is in some of the barnacles (discovered by Darwin), in which the big individuals are of double sex (hermaphrodite). Adhering to the shells of these are found minute dot-like "supplemental males." It is to be observed that these are instances where the inferiority of the male is an obvious measurable fact. In the mammals, the group of vertebrate animals to which man belongs, the male possesses measurably greater activity and size than does the female, and is provided with more powerful natural weapons, such as teeth and horns. He entirely dominates and controls the female, or a whole company of females, and in no case is there equality of the sexes, or any approach to it, still less inferiority of the male. It is, perhaps, a question whether "by taking thought" this natural inferiority of the mammalian female can be changed.

The survival of Rotifers, especially of a pink-coloured species (called Philodina roseola), after long drying or "desiccation," has been experimentally studied. It is found that if the water in which some are swimming is placed in a watch-glass and allowed to dry up rapidly the Rotifers are killed, none reappear when after a few hours fresh water is poured into the watch-glass. But if a few grains of sand or particles of moss are present from the first in the water the final drying up takes place more slowly and the Rotifers find their way between the sheltering fragments, where the water remains long enough to give them time to form a little gelatinous case, each for itself. When thus encased they survive, motionless, for months. The experiment has often been made, and is not in doubt. According to trustworthy statements, Philodina can thus survive even for so long as five years. The processes of life are arrested, but the drying has not proceeded to the extent which is called chemical drying or dehydration. The tiny Rotifers are still of soft consistence: the protoplasm is not chemically destroyed. When one is watched with the microscope as water is allowed to flow round it after several months of dust-like aridity, it is seen to emerge from its protective case and at once to commence swimming and searching for food by means of the currents directed towards its mouth by its so-called "wheel-apparatus." I may just say that in the case of the slime-mould called "flowers of tan" the protoplasm dries to the consistency of hard wax, and I have kept it for years in that state and then revived it by moisture into full activity and growth. I used also at one time to keep in my laboratory a supply of the dried yellow lichen from apple-trees, in which one could always rely upon finding the animalcules called "Macrobiotus" or "water-bears" ready to be revived from a desiccated condition, after three or four years passed in that condition.

Many of the Rotifers carry their eggs when ripe extruded from the body in two bunches or clusters, as is the habit also of the little microscopic shrimps known as Cyclops. There is a whole group of Rotifers which fix themselves by the tail, when full grown, to some solid support. Each then forms a protective tube or case around itself, from the mouth of which it puts forth its wheel-apparatus and into which it can retire for protection. Some of the largest and most beautiful of the wheel animalcules belong to this group of fixed or sedentary Rotifers. The crown animalcule (Stephanoceros) is one of these, having what are discs edged with vibrating hairs in the common Rotifer—here drawn out into a circlet of tapering lobes like the points of a coronet (Fig. 37 (*bis*), B). Another is the floscule (Floscularia), in which the wheel-apparatus has the form of five knobs arranged on a pentagonal disc around the mouth (A in same figure). Each knob has a bundle of excessively fine, long, stiff, motionless hairs spreading out from it ready to entangle food particles which may drift into contact with them. I used to find the stems of the fresh-water polyp (Cordylophora) of Victoria Dock a sure source of supply of these fine little creatures. When seen under the microscope as brightly illuminated glassy florets on a black ground (by what is called "dark ground illumination") their strange delicacy and beauty cannot be surpassed. A rare species of floscule (which I have never seen) has extra-long and fine filaments, each of which shows a fine streaming current in its substance, and is, in fact, a naked filament of living protoplasm like one of the ray-like filaments of the sun-animalcules.

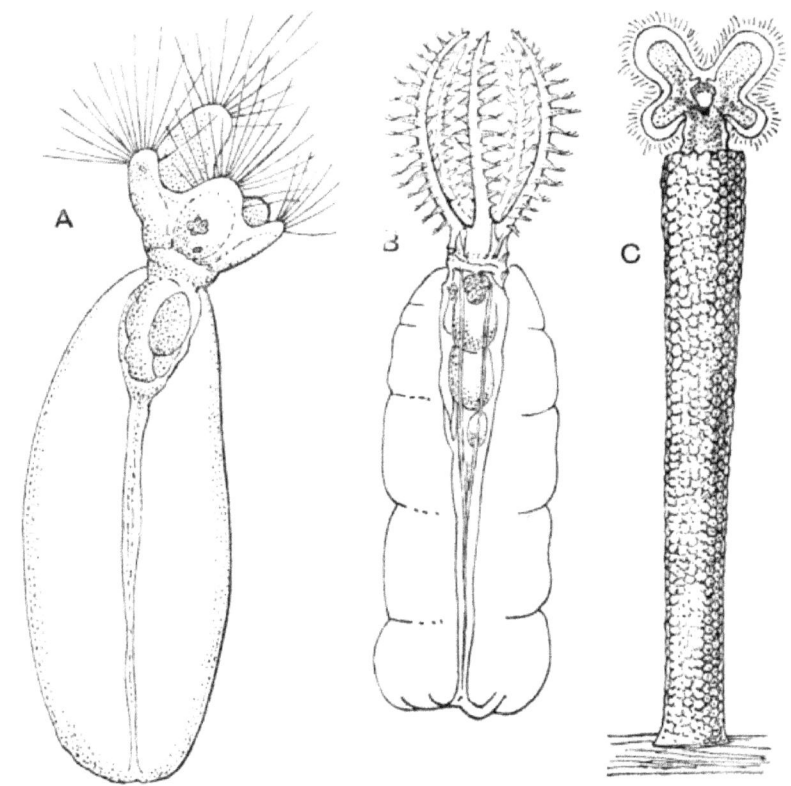

FIG. 37 (*bis*).—Three tube-building wheel animalcules. *A*, Floscularia campanulata. *B*, Stephanoceros Eichhornii. *C*, Melicerta ringens.

The most curious of the tube-building Rotifers are those which form their tubes of little, equal-sized pellets of solid matter—as it were, "bricks"—which they first form by compacting fine particles in a special pit on the head and then build them up and cement them together in rows to form the tube, adding row after row as the animal itself increases in length (Fig. 37 (*bis*), C). These are known as Melicerta; and, though some kinds use any minute particles to make their bricks, one kind is frequent which uses its own excrement for this purpose. By feeding the little creatures first with food coloured with carmine and then with blue-stained material, one can obtain alternate rows of pink and blue pellets, carefully manufactured and laid in position to build up the growing length of tube.

Melicerta has certainly an extraordinary and economical way of disposing of that refuse which we larger creatures carefully remove from our habitations and should be very unwilling to employ as building material. The individuals of one rare and interesting kind of the tube-builders, after swimming freely in the youngest stage, settle down together and form their gelatinous transparent tubes side by side, to the number of fifty or more, in such a way as to produce a perfect sphere, a twentieth of an inch or more in diameter, built up of fused jelly-like tubes radiating from a common centre. The inhabitant of each tube is quite separate from and independent of his neighbours, but they all protrude their vibrating wheel-apparatus simultaneously, and cause the glass-like ball to rotate and travel through the water. Many years ago I found this beautiful little thing in a small moss-pool (not more than 3 ft. wide), high up the sloping-side of the north-west section of Hampstead heath, above the "Leg of Mutton Pond." The well-meant care of the public guardians of the heath has now drained this region, and my little moss-pools and the "bog," in which grew the Drosera, or Sun-dew, and the Bog-bean and such plants, have gone for ever. But we must console ourselves with the fact that the same progressive expansion of the great city has given us electric railways, tubes, and tramways by which we can go farther afield than Hampstead in a few minutes, and still find moss-pools and the undisturbed glories of ancient swamps and bog-land.

Many of the Rotifers have a pair of ruby-red eyes, and in some of them there is a minute crystalline lens overlying the red sensitive spot, which receives the fibres of the optic nerve coming from the brain—one on each side. It is almost incredible that so minute a creature—often only the one-fiftieth of an inch long when full grown—should have a nervous system and special organs of touch (sensory hairs) as well as eyes, and on the other hand muscles running from one attachment to another and called into activity by nerves connected with this same central brain. The pair of branched tubes, which end internally in flickering "flame-cells" and open externally far back at the vent, are kidneys. Similar tubes called "nephridia" or little kidneys are found in many of the smaller animals; the earthworm has a pair in each ring of its body.

There is little doubt that the wheel animalcules are related in pedigree to the primitive ancestors of the marine segmented or annulate worms, which also gave rise to the ringed leg-bearing jaw-footed creatures with hard skin, called Crustacea, Arachnids and Insects (the Arthropods). The wheel-apparatus or cilia-fringed discs of the Rotifer is seen in the young stages of many marine worms, and also in the young of marine snails, known as the "veliger"— "velum" or "sail" being the name given to the wheel-apparatus of the young snails (see the drawing on p. 181). There are very minute marine annulate or segmented worms (Dinophilus and others), which come near to the Rotifers in many features, whilst the ringed or segmented character of the body is obvious in the common wheel animalcule.

The Rotifers are so small that they are built up of very few "cells" or nucleated units of protoplasm. Many of them are of smaller size than some of the big infusorian animalcules, which consist of a single cell. The Rotifers are probably a dwindled pygmy race descended from ancestors of ten or a hundred times their linear measurement. It is an important fact that in the possession of a toothed gizzard, in the hard body-case or cuirass of some kinds, and in Pedalion's rapidly-moving legs or paddles, fringed with plumose hairs and moved by that peculiar variety of muscular tissue which is called "striped muscular tissue," the wheel animalcules give evidence of relationship to the Crustacea—that is to say, it appears to be probable that they were derived from the common ancestor of marine worms and Crustacea before those two lines of descent had diverged.

Rotifera or wheel animalcules are found all over the world, in the tropics, the temperate zones, the Arctic and Antarctic, and many species have a world-wide distribution. They occur in fresh waters and in the sea, in great lakes, in gutters which dry up, in pools in the polar regions and on high mountains which are solid ice for the greater part of the year. A few are parasitic, some living on the legs of minute Crustacea. One which I discovered in 1868 in the Channel Islands lives in crowds on the skin of a remarkable sea-worm (Synapta), which burrows in the sand, exposed at low tide. It holds on (as I found and figured) by a true sucker, which replaces the forked tail of other commoner Rotifers. It was named "Discopus" by Zelinka, who searched for it in consequence of my description, and gave a very detailed account of it. Others are parasitic inside earthworms, and one is found inside the globe animalcule Volvox! Another causes the growth of warts or "galls" in a curious kind of Alga called Vaucheria.

CHAPTER XV

SUSPENDED ANIMATION

OUR leading newspapers, with rare exceptions, never report the discoveries announced at our scientific societies. But they often seek to astonish their readers with silly stories of monsters said to have been seen in tropical forests, ghostly "manifestations" and such rubbish transmitted to them at a high price by crafty "newsmongers," and do much harm to themselves and to the public thereby. On the other hand, foreign newspapers do occasionally report the proceedings of their local Academies—and then "our own correspondent" telegraphs to London with a flourish, a confused report of what he has read and ignorantly imagines to be "a startling discovery" because he knows nothing whatever of the subject. Thus shortly before the recent war—the confirmation by a French experimenter of the fact, long since demonstrated, that the seeds of plants can survive exposure to very low temperature, was announced with ridiculous emphasis by one of these "fat boys" of journalism *pour épater le bourgeois*.

A temperature very near to that of the total absence of that molecular movement or vibration which we call "heat," can now be attained by the use of liquid hydrogen, which enables us, by its evaporation, to come within a few degrees (actually three!) of that condition known as the "absolute zero." We divide into one hundred equal steps or degrees the column of liquid (mercury, spirit, or other liquid) of a thermometer as it expands from the shrunken bulk which it occupies when placed in freezing water to the full length which it attains when the water is heated to boiling point. This is called the centigrade scale, or scale of a hundred degrees. But, as we know by the records of travellers in the Arctic regions and by the experiments made in laboratories, there are "degrees" of coldness or diminution of heat which are much below that of freezing water, and can be measured by the further shrinking of the column of liquid in the thermometer, so that we record "degrees below zero centigrade," each of the same length as those above it and corresponding to the same "quantum" of decrease or increment of heat. As we pass from the temperature at which water is solid to that much lower or diminished state of hotness at which mercury becomes solid, the shrinking column of the thermometer (in which a liquid is used not rendered solid by this amount of cooling) falls through 39 degrees of the centigrade size, so that we say that mercury freezes at minus 39 or at 39 degrees below zero of the centigrade scale. The conclusion has now been reached that the absolute zero or cessation of all heat in a body is represented by a fall of no less than 273 degrees below zero on the centigrade scale. Hydrogen gas becomes a liquid at 252 degrees below zero centigrade, and a solid at 264 degrees. If we start our counting of those degrees or increments of heat, of which there are 100 between the freezing and boiling points of water, at the absolute zero or condition of total absence of heat, we must say that hydrogen "melts"—that is, passes from the solid to the liquid state—at 11 degrees (absolute), and boils at about 20 degrees (absolute), whilst water does not melt until 273 degrees (absolute) of temperature are reached, and boils at 373 degrees above the absolute zero.

It is the fact that, from the year 1860 onward, numerous observers have experimented on the influence of very low temperatures upon seeds, and have uniformly shown that the power of germination and healthy growth of the seeds is not destroyed by exposure to very low temperatures. The celebrated Swiss botanist, De Candolle, published the first careful observations on this subject in conjunction with Raoul Pictet, who had devised an apparatus for producing exceedingly low temperatures. Pictet in 1893 exposed various bacteria and also seeds to a temperature of nearly 200 degrees below zero centigrade without injury to them. They "resumed" their life when gradually restored to the normal temperature. Pictet concluded that since all chemical action of the kind which goes on in living things requires a certain degree of temperature for its occurrence, and that this is demonstrably considerably higher than minus 100 degrees centigrade, we must suppose that all chemical action in living things (as in nearly all other bodies) is annihilated at 100 degrees below zero centigrade. Accordingly he maintained that what we call "life," or "living," is a manifestation of chemical forces similar to those shown in other natural bodies, and liable to interruption and resumption by the operation of unfavourable or favourable conditions as are other chemical processes. In 1897, Mr. Horace Brown and Mr. F. Escombe published, in the Proceedings of the Royal Society of London, an account of experiments in which they exposed seeds of twelve plants belonging to widely different natural orders to a temperature varying from 183 degrees to 192 degrees below zero centigrade for a period of 110 consecutive hours (about four days and a half). As a result the germinative powers of the seeds showed no appreciable difference from that of seed not subjected to cold, and they produced healthy plants. The low temperature was obtained by the use of liquid air in a vacuum-jacketed flask (like the well-known "thermos" flasks), into which the seeds were introduced in thin glass tubes. Professor M'Kendrick had previously shown that the putrescence of meat, blood and milk by bacteria infesting them was temporarily arrested, but not permanently so, by exposing those substances for one hour to a temperature of

182 degrees below zero centigrade. It appeared that the putrefactive bacteria present in those substances were not destroyed by that degree of cold, but returned to a state of activity when the normal temperature was restored. Professor M'Kendrick also showed that seeds would germinate after exposure to like treatment.

All this is ancient history, twenty years and more in the past. The experiments of a French observer, mentioned at the beginning of this chapter as foolishly trumpeted in a London paper, were of service as confirming the extensive and careful work of his predecessors. It is only when our old well-bottled discoveries have, however tardily, been brought before the Paris Academy of Sciences and sent back to us by the Paris correspondents of news agencies as "startling novelties" and "amazing discoveries" (twenty years old), that any attempt is made to mention them in the London daily Press. And then they are announced without any reference to their true history. This habit of culling stale morsels of information from the proceedings of foreign academies points to the fact that there is incompetence both in the purveyor and publisher of such scraps. If our newspaper editors must publish scraps about scientific novelties, they should employ educated assistants to see that they do not make themselves ridiculous. The scraps which come round to our newspapers from Paris are usually plagiarized from a French newspaper by some one who has a very imperfect knowledge of the subject to which they refer, and adds his own blunders to those of the original reporter.

The action of extreme cold in arresting life in such minute organisms as plant seeds and bacteria without destroying the possibility of the resumption of those chemical and physical changes when warmth is restored, is dependent on the fact that those chemical changes can only proceed in and by the aid of liquid water. When thoroughly frozen the chemical constituents of minute organisms and seeds—which until frozen were living and undergoing continuous, though perhaps slow, change—become solid, and can no longer act on one another or be acted on by surrounding chemical bodies equally reduced in temperature. They may be compared to the solid dry constituents of a Seidlitz powder—one an acid, the other a carbonate. So long as they are dry they remain—when mixed and shaken together—inert, without action on one another. Even if one is dissolved in water and then frozen solid and mixed in a powdered state with the other at an equally low temperature the mixture remains dry and inert. Nothing happens so long as the low temperature is maintained. But if we raise the temperature above the freezing-point—so as to liquefy the solution—chemical action will immediately ensue. With much fizzing and escape of gas the two chemicals will unite. The effect of cold on living matter is of this nature. It is a real "suspension" of the changes which were—however slowly and quietly—going on before complete solidification of the protoplasm by freezing. A frozen seed and frozen bacteria are in a state of "suspended animation."

It is not the fact that absolutely all chemical union and change whatsoever is prevented—that is to say, arrested or suspended—by extreme cold, although the union with oxygen and other such changes of the essential material of living things, which we call "protoplasm," and most other chemical changes are thus arrested or suspended. The most striking exception is that of the most active of all elements, the gas fluorine, which becomes a liquid at 210 degrees below zero centigrade, and in that condition attacks turpentine if brought into contact with it at the same low temperature with explosive force. Even solid fluorine combines with liquid hydrogen with violent explosion. It seems certain, however, that elements or chemical compounds brought into the solid (not merely liquid) condition by extreme cold cannot act chemically upon other bodies in the same solid condition, even when they would at normal temperatures so act with the greatest readiness, because they are then either liquid or gaseous.

The conception of an arrest of the changes in organisms, which we call life, followed by their resumption after a greater or less interval of suspense, was long ago suggested and discussed before we had knowledge of the action of low temperatures. The winter-sleep of some animals and the "comatose" condition sometimes exhibited by human beings had led to the notion of "suspended animation." But a careful study of hybernating animals and of human instances of prolonged "coma" satisfied physiologists nearly 100 years ago that the processes of life—the beating of the heart and the respiration—were not actually and absolutely suspended in these cases, but reduced to a minimum. The chemical processes connected with life were still very slowly carried on.

Again, a great deal of interest and discussion was excited in the last century by the drying up of delicate yet complex aquatic animalcules, such as the Rotifers (the wheel animalcules described in our last chapter) and Tardigrades (bear animalcules), and the fact that after their preservation as mere dust for many months dried on a glass-slip they could be revived and made to return to life by wetting them with a minute drop of water, whilst the whole process of revival was watched under the microscope. Letters were published in the "Times" in the "fifties" by the Rev. Lord Sydney Godolphin Osborn, describing his observations and experiments on these animalcules.

The yellow slime-fungus called "flowers of tan," after creeping as a naked network of protoplasm over the "spent tan," thrown out from tan-pits, will in dry weather gather itself into little knobs, each of which is as hard and brittle as a piece of sealing-wax. Yet (as I have repeatedly experienced in using material given to me by the great botanist, de Bary) a fragment of one of these hard pieces, if carefully guarded in a dry pill-box for two or three years, will when placed on a film of water at summer-heat gradually absorb moisture and expand itself into threads of creeping, flowing protoplasm, nourish itself, and grow and reproduce. It was formerly suggested in regard to these cases of resuscitation after drying, as also in the case of seeds which germinate after being kept in a dry condition for many years, that really they were not thoroughly dried, but were sufficiently moist to allow of very slow oxidation and gas exchange, which it was said was so small in amount as to escape observation. There was a plausible comparison of the condition of these dried organisms to that of hybernating mammals, desiccated snails, and comatose men. It was held that here, too, the life-processes were not absolutely arrested, but reduced to an imperceptible minimum.

This view of the matter was connected, no doubt, with a traditional assumption that life was an entity—an "anima animans"—which entered a living body, kept it continually "going" or "living," and if driven out from it could not return. Curiously enough, Mr. Herbert Spencer seems to have been (perhaps unconsciously) affected by this traditional view, since he defined life as "the continuous"—that is the important word—"adaptation of internal to external relations." This definition prejudiced the view of some distinguished physiologists on the question of "suspended animation," and I remember a very warm dinner-table discussion with Michael Foster and other friends, some twenty-five years ago, when I put forward the view that so long as the intimate structure—in fact, the chemical structure—of the protoplasm of a living thing is not destroyed, it does not "die" though all chemical change in it may be arrested. I compared the dried seed and dried animalcule—as I would now compare the frozen seed and the frozen bacteria—to a well wound watch which is stopped by the intrusion of a needle between the spokes of its balance wheel, or, better, by the cooling on the wheel of a tiny drop of soft wax so as to clog it. The works of the watch are rendered absolutely motionless, but it is not "dead." As soon as the needle is removed or the tiny speck of wax melted by a gentle warmth it resumes its movement. It is, as we say, "alive again." So, too, the frozen or dried organism is absolutely motionless. No chemical movements can go on in it. They are stopped by the solidity set up by freezing, or in the case of simple "desiccation," by the absence of the moisture necessary for bringing the chemical molecules into contact. If protected from destructive agents, the mechanism remains perfect for just so many years or centuries as that protection lasts. Whenever the frozen organism thaws or the dried organism becomes wet, the life-processes are resumed, the seed germinates, the bacteria grow and multiply.

Thus we see what are some of the points of interest and importance raised by the old experiments of Pictet, M'Kendrick, and Horace Brown, the results of which were the same as those announced as Parisian novelties. I have yet to say a few words as to the reason why we cannot produce "suspended animation" in higher organisms or in man by the application to them of extreme cold. Further, the influence of extreme cold on the possible passage, through space, of living germs from other worlds to this earth—a possibility suggested by the late Lord Kelvin—requires some consideration in connection with the striking experiments with phosphorescent bacteria described ten years ago by Sir James Dewar.

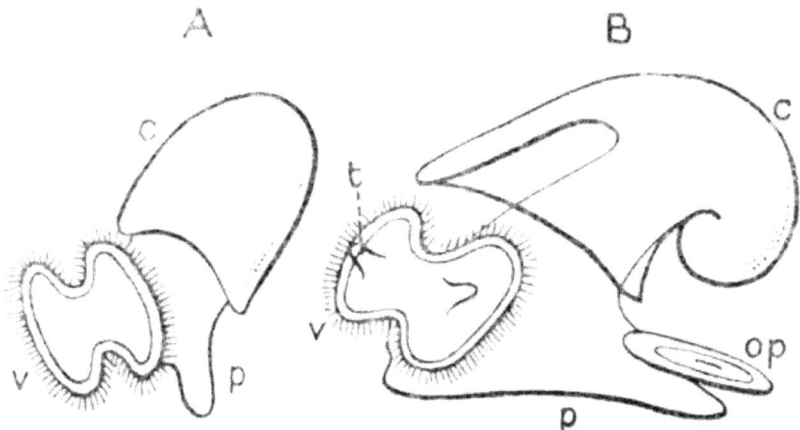

Young stages of growth or Veliger larvæ of marine snails, showing the ciliated band or velum which is identical with the wheel apparatus of the Rotifers or Wheel animalcules.

CHAPTER XVI

MORE ABOUT SUSPENDED ANIMATION

I GAVE some account in the last chapter of the experiments made within the last twenty years, which have shown that, in certain very simple organisms and in seeds, all chemical change can be stopped by the application to them of methods of freezing. The continuous changes which go on in these living things under ordinary circumstances are arrested by the solidification of what was more or less "moist" material. Water in the liquid state, though it may be in extremely minute quantity, is necessary for the chemical combinations and decompositions which go on in living things. Hence not only the solidification of all moisture, in or having access to the living bodies experimented on, arrests those chemical combinations and decompositions, but very thorough drying also has this result. Yet either on thawing the frozen liquid or supplying moisture to the "dried up" organism, the previously continuous chemical and physical changes are resumed as though no arrest or suspension of them had occurred. No limit is known to the length of time during which this arrest may be continued, and yet a resumption of living changes occur when the cause of arrest—namely, either solidification by cold or else dryness—is removed. The apparatus—the exact structure and the exact chemical materials—of the seeds or the bacteria remains uninjured and unchanged by either freezing or drying carefully applied. It is, of course, impossible to guarantee that no accident, no unforeseen change in the surroundings, shall take place and destroy in one way or another the experiment. But the arrest of all change, such as goes on in life, has been, in many experiments, maintained under careful supervision and protection for several months, and yet life has been resumed when the cause arresting chemical change has been removed. The presumption, then, is in favour of the possibility of the arrest being maintained for an unlimited period, and yet at any time being resumed when the arresting cause (cold or dryness) is removed.

Before what we may call "the suspensory action" of very low temperatures had become generally known, the question occurred as to whether seeds kept in a dry condition for several months, or even years, and yet capable of germination when placed in moist earth, are during their dry condition undergoing any chemical changes. The matter presented itself in this way. The dry seeds can germinate when sown, therefore they are not dead, but living. According to various physiologists and philosophers (*e.g.*, Herbert Spencer), life is a continuous adjustment of internal to external relations. Burdon Sanderson, the Oxford professor of Physiology, declared that "life is a state of ceaseless change." If this is a correct conception, and if by "living" we mean, as the great Oxford English Dictionary tells us, "manifesting the property called life," then the seeds which, though dry, are still "living" or "alive" or "endowed with life," should yield some evidence of the "ceaseless change" (by which is meant chemical change) of which, as things not dead but living, they are supposed to be the seat. The late Dr. George Romanes published in 1893 some experiments on this matter. We know that free oxygen is very generally (though not universally) necessary for the continual chemical changes which the minutest as well as the largest plants and animals carry on. Romanes enclosed a quantity of dry seeds in glass tubes, from which he pumped out all gas as completely as possible—that is to say, all except one-millionth of the original volume. He also expelled all oxygen by replacing it by other gases. As a result of this treatment, continued for as much as fifteen months, he found that neither a high vacuum nor subsequent exposure for twelve months in separate instances to oxygen or to hydrogen, or nitrogen, or carbon monoxide, or carbon dioxide, or hydrogen sulphide, or the vapour of ether or of chloroform, had any effect on the subsequent germinative power of the seeds employed. These experiments proved that anything like respiration by ordinary gaseous exchange with the atmosphere was *not* going on in the seeds, and that if they are the seat of "ceaseless change" because not dead, the changes must be chemical interactions of some kind or other within their protoplasm.

The keeping of seeds and also of bacteria for days and even months—at temperatures as low as 100 degrees below zero centigrade—and their subsequent resumption of life, has removed the possibility (not excluded by Romanes) of the occurrence of chemical interactions within the substance of these organisms preserved during long periods of time, and yet not ceasing to be what is ordinarily called "alive," or endowed with "life." It is time that we should definitely abandon Herbert Spencer's and Burdon Sanderson's definitions or verbal characterizations of "life." The word "life" is commonly and properly used to designate the condition of a "living thing" or a thing which is "alive." A thing which has lost life—that is, which was living, but is so no more, and cannot be "restored to life" or resuscitated—is, in correct English, said to have "died," or to be "dead." The motionless, unchanging frozen seed or bacterium, which resumes its living activities when carefully thawed, has *not* "died." The mere fact that it can be resuscitated justifies the application to it, according to correct English usage, of the word "alive"—it is still "alive." It is not possible to alter the significance of the words "life," "living" and "alive," so as to retain the definitions of Herbert Spencer and Burdon Sanderson as correct. They are incorrect. Life is not continuous or ceaseless change. It is a property of the more active substance of plants and animals which has special structure and definite chemical constituents. The property is, no doubt, usually manifested under normal conditions of temperature, light, moisture, pressure, chemical and electrical surroundings in a continuous series of changes, both chemical and physical. But at exceptionally low temperature, and in other arresting circumstances these changes can, in a few exceptional organisms, be absolutely stopped, though the organism in which the changes cease is uninjured as a mechanism. It still possesses "the property of life"—is still "alive" although motionless and unchanging. Its life is in suspense, as is that of a clock with arrested pendulum.

The unjustified conception of "life," or "living," or being "alive," and not dead, as necessarily a state of incessant chemical and other change, is bound up with the old fancy that life is not to be considered as a state or motion of a special and complex structure called protoplasm, but is a thing, a spirit or an essence, which takes possession of organic bodies and makes them "live." According to Sir Oliver Lodge, if chemists could build up the chemical materials which constitute protoplasm, the protoplasm so made by them would not live. It would (he stated at the meeting of the British Association in Birmingham in 1912) have to receive a charge or infusion, as it were, of this thing suggested by the word "life." It cannot live itself (according to the suppositions of Sir Oliver), but serves as the vehicle, the receptacle, for this supposed intangible entity "life." In the same imaginative vein, our grandfathers used to say that heat was due to the entity or "fairy" "caloric" which could be enticed into or driven from material bodies, making them "hot" by its presence and cold by its greater or less exclusion. The suspended animation of frozen germs and their return to life when warmed could thus be represented as an affection or affinity between the fairy "Vitalis" and the fairy "Caloric," the former fleeing from the body and waiting near when the latter deserts his place, but returning to happy union with "Caloric" when he again, however feebly, pervades once more the vehicle provided for "Vitalis." Such imaginary essences are not of any assistance to us in arriving at a knowledge of the facts, and so far from helping us to a comprehension of the ultimate nature of things (which we have no reason to suppose that it is possible for us to attain) their introduction tends to the substitution of imaginary causes and unverified assumptions for the carefully-tested and demonstrated conclusions of science.

In 1871 Lord Kelvin, when president of the British Association, suggested that the origin of life as we know it may have been extra-terrestrial, and due to the "moss-grown fragments from the ruins of another world," which reached the earth as meteorites. It was objected to this that the extreme cold—very near to the absolute zero—which prevails in interstellar space would be fatal to all germs of life carried by such meteoric stones. But twenty years later Sir James Dewar showed that this objection did not hold, since at any rate some forms of life—certain bacteria—could survive an exposure of several days to a temperature approaching the absolute zero. Later Sir James made some very striking experiments by exposing cultivations of phosphorescent bacteria to the temperature of liquid hydrogen (252 degrees below zero centigrade). These bacteria may be obtained by selective cultivation from sea-water taken on the coast, in which a few are always scattered. A rich growth of these bacteria in gelatine broth gives off a brilliant greenish light when shaken with atmospheric air or otherwise exposed to oxygen. The light is so intense that a glass flask holding a pint of the cultivation gives off sufficient light to enable one to read in an otherwise dark room. The emission of light is dependent on the chemical activity of the living bacteria in the presence of oxygen. In the absence of free oxygen they cease to be luminous. As soon as they are killed the light ceases. When they are frozen solid the light ceases, even in the presence of free oxygen gas. When a film consisting of such a culture is frozen solid it will remain inactive if the low temperature be maintained for months, though exposed to free oxygen gas, and then, as soon as it is liquefied by a gentle rise in temperature, the active changes recommence, and the phosphorescent light beams forth. Sir James Dewar exposed such films to the low temperature of liquid hydrogen for (so far as I remember) six months, and obtained from them at once the evidence of their living chemical activity, namely, their "phosphorescence," as soon as they were thawed. In the frozen state, at a temperature of minus 250 degrees centigrade, nothing, it appeared, could injure these phosphorescent bacteria. No chemical can "get at them" at that temperature, the most biting acid, the most caustic alkali cannot touch them when, like them, it is in a hard, solid condition. Powdering the film by mechanical pressure has no effect on the bacteria. They are too small to be crushed by any mill. Such germs would, it seemed, surely be able to travel through interstellar space, as suggested by Kelvin.

Then it occurred to Sir James that light—the strangely active ultra-violet rays of light—might be able to disintegrate and destroy the bacteria, even when frozen solid at the lowest temperature. He exposed his frozen cultures to strong light, excluding any heat-giving rays, and found that the bacteria no longer recovered when subsequently the culture was thawed. Light, certain rays of light, can, it thus appears, penetrate and cause destructive vibrations in chemical bodies protected from all other disintegrating agencies by the solidity conferred by extreme cold. I am not able to say, at the moment, how far this important matter has been pursued by further experiment, nor whether what are called the "chemically active" rays of light and other rays such as the Röntgen rays can effect chemical change in other bodies (besides living germs), upon which they act at normal temperatures, but in regard to which they might be expected to be inoperative when the bodies in question are brought into the peculiar state of inactivity produced by extreme cold. Since light is far more intense outside our atmosphere than within it, it seemed, at first, that the demonstration of its destructive action on frozen germs puts an end to Kelvin's theory of a meteoric origin of life. It must, however, be remembered that minute living germs could conceivably be protected from the access of light by being embedded in even very small opaque particles of meteoric clay. So Lord Kelvin's suggestion as to the travelling of life on meteoric dust cannot be set aside as involving the supposition of the persistence of life in conditions known to be destructive of it.

The great interest in former times in relation to "suspended animation" has naturally been in relation to the occurrence of this condition in man and the possibility of producing it in man by this or that treatment. There is no doubt whatever, at the present day, that "death-like" trances, whether occurring naturally or after the administration of drugs, in the case of man and of higher animals, are not due to that complete suspension of living changes which we can produce, as I have here related, in certain lower forms of life. These death-like trances are merely cases of reduction of the living changes to a very low degree. [6]

The bodies of all but the simplest animals and plants are too large and too complex to survive the bursting and disruptive action of extreme cold, due to the unequal distribution of water within them and its irresistible expansion when frozen. Their living mechanism is broken, mechanically destroyed by this expansion. We cannot hope to apply cold to man so as to produce "suspended animation." It is true that experiments are on record in which fish and even frogs have survived enclosure within a solid mass of ice by the freezing of the water in which they were living. But careful experiments are wanting which would demonstrate that these animals were actually frozen through and through, and that either fish or other cold-blooded animals can survive a thorough solidification by freezing of their entire substance. Such survival cannot be pronounced to be impossible, but it has not been demonstrated in any cold-blooded animal—even shell-fish or worm or polyp—let alone a warm-blooded mammal. It appears that, apart from disruptive effects, the protoplasm of even very minute and simple organisms, such as the Protozoa, does not in all kinds, even if in any, survive exposure to great cold. The toleration of great cold and return to living activity after thorough freezing is, it appears, a special quality attained by the living material of vegetable seeds and by many kinds of bacteria. A similar special toleration of high temperatures, a good deal short of the boiling point, but high enough to kill most plants and animals, is known to characterize certain bacteria and allied Schizophyta found in hot springs. It is a matter of common knowledge that many animals and plants are killed by a temperature (whether too high or too low for them) which allows others to flourish and may be necessary for their life. Minute organisms (flagellate monads) have been cultivated experimentally in a nourishing liquid, the temperature of which was raised daily by one or two degrees until the liquid was so hot that the same species of organism was at once killed by it when abruptly transferred to it from liquid at ordinary summer temperature.

The true "suspended animation" of many vegetable seeds and of many kinds of bacteria under the influence of cold is not an exhibition of a general property of living things, but is due to a special quality of resistance gradually attained by natural selection of variations a little more tolerant of cold or of drought than are the majority. It is of life-saving value and a cause of survival to the species of plants and bacteria concerned. No doubt there is need of further experiment on the subject of the "killing" or destructive effect exerted by different degrees of diminution of temperature upon the protoplasm of all kinds of organisms, and with the knowledge so obtained we shall be able to frame a conception of the actual mechanical and chemical peculiarities of the protoplasm of those bacteria and of those vegetable seeds which can be exposed to the extreme of cold for many months or for an indefinite period and yet subsequently recover or live again. Probably in order to survive freezing, protoplasm must be, not absolutely dry, but free from all but a minimum of moisture.

FOOTNOTE:

[6] See the chapter on "Sleep," in my "Science from an Easy Chair," Methuen, 1909.

CHAPTER XVII

THE SWASTIKA

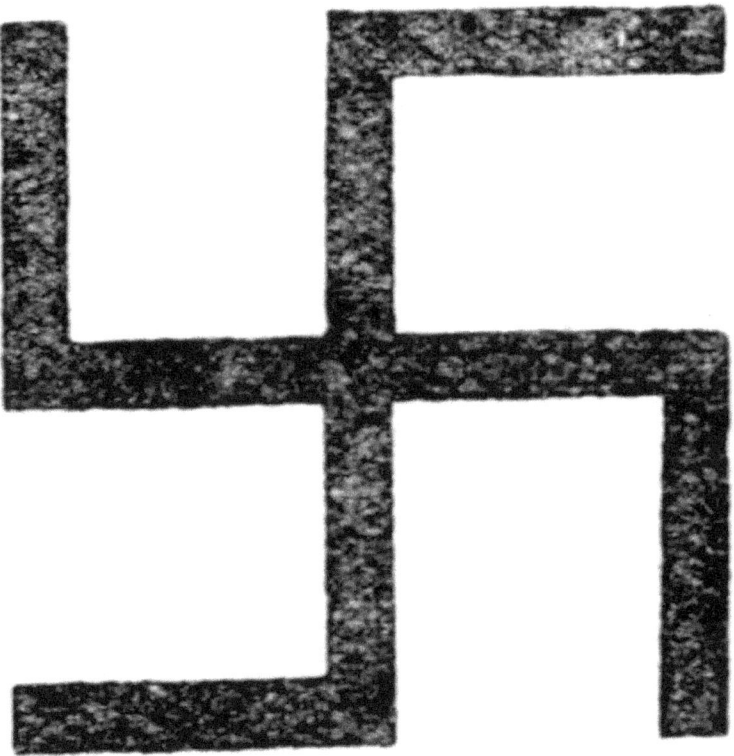

FIG. 38.—The swastika in its simplest rectangular form. It may turn
to the right, as here, or to the left, a less usual thing, but without
significance.

A GOOD many people have never heard of the Swastika.
It is an emblem or device such as is the Cross or the
Crescent. Here it is (Fig. 38) in its most simple and most
common form. In India it is in common use at the present
day, and has been so for ages. It is the emblem of good
luck. The name "Swastika," by which it is widely known,
is a Sanskrit word meaning "good luck." The word is
composed of Su, the equivalent of the Greek *eu*, signifying

"well" or "good," and asti (like the Greek *esto*), signifying "being," whilst ka is a suffix completing the word as a substantive. The sign or emblem called Swastika must have existed and been largely used in decoration of temples, images, swords, banners, utensils, and personal trinkets of all sorts long before this name was given to it. It has a name in many widely separate languages. It is often referred to by English writers as the fylfot, the gammadion, and the "crux ansata," also as the "croix gammée." It is often found more roughly drawn (on pottery or cloth) as shown in Fig. 39. Often the arms of the cross are bent rigidly at right angles as in Fig. 38, but they are often only curved as seen in Fig. 39, C, or curled spirally as in B, when it is called an "ogee." The arms of the Swastika are sometimes bent to the right as in Fig. 38, and sometimes to the left as in Fig. 39. This difference does not appear to have any symbolic significance, but to depend on the fancy of the artist.

FIG. 39.—Three simple varieties of the swastika. *A*, the normal rectangular. *B*, the ogee variety (with spiral extremities). *C*, the curvilinear or "current" variety.

FIG. 40.—Footprint of the Buddha, from an ancient Indian carving, showing several swastikas. (Fergusson and Schliemann.)

In Figs. 40 to 45 a few examples are shown of the Swastika from various places and ages. It was in use in Japan in ancient times, and is still common there and in Korea. In China, where it is called "wan," it was at one time used, when enclosed in a circle, as a character or pictograph to signify the sun. It has been employed in China from time immemorial to mark sacred or specially honoured works of art, buildings, porcelain, pictures, robes, and is sometimes tattooed on the hands, arms, or breast. In India it is widely used in decoration by both

Buddhists and Brahmins; children have it painted on their shaven heads, and it is introduced in various ceremonies. The gigantic carved footprints of Buddha from an Indian temple drawn in Fig. 40 show several Swastikas on the soles of the feet and on the toes. In the Near East and in Europe the Swastika is no longer in use: it is not, in fact, popularly known. But in ancient and very remote times it was in constant use in these regions, especially by the Mykenæan people and those who came under their influence, and also by the people of the Bronze Age—before the use of iron in Europe. Fig. 41 shows a vase of Mykenæan age (about 1200 years B.C.) from Cyprus ornamented with Swastikas. Hundreds of terra-cotta "spindle-whorls" like Fig. 42 were found by Schliemann in excavating Hissarlik and the site of ancient Troy, and some of them date from 3000 B.C. in layers of different ages. The vase on which is painted the ornament shown in Fig. 43 is from Bœotia, and belongs to the same early period—the "Mykenæan" or "Ægæan" before that of the Hellenes. It still survives in the pottery of the Dipylon period (*circa* 800 B.C.), as is seen in the fragment drawn in Fig. 6, Chapter I. The later Greeks of the great classical period (Hellenes) did not use the Swastika. Nor has it been found on the works of art of the ancient Egyptians, nor in the remains of Babylonia, Assyria or Persia. It, in fact, seems to have belonged especially to that ancient "Minoan" civilization, the remains of which are found in Crete and the other Greek islands. The same culture and the same race is revealed to us by the discoveries of Schliemann at Mykenæ and other spots in Greece, and at Hissarlik, the seat of ancient Troy. The Mykenæan art seems not to have been transmitted to the post-Homeric Greeks, nor to Egypt, nor to Babylonia and Assyria. The Swastika seems, like the "flying gallop" of Mykenæan art, to have travelled in very ancient times by a north-eastern route to the Far East.

I have given some account of the latter, with illustrations, in "Science from an Easy Chair," Second series. Like the representation of the galloping horse, with both fore and hind legs stretched and the hoofs of the hind legs turned upwards, the Swastika is found in the remarkable metal work (Fig. 43 *bis*) discovered in the necropolis of Koban, in the Caucasus, dating from 500 B.C. The Swastika and the "flying gallop" probably travelled together across Asia to China and the Far East, and so eventually to India on the one hand and Japan on the other—the Swastika thus escaping altogether, as does the pose of the "flying gallop," the Near East and later Greece. This is a very remarkable and interesting association.

FIG. 41.—Vase from Cyprus (Mykenæan Age, *circa* 1200 B.C.); painted with lotus, bird and four swastikas (Metropolitan Museum, New York City).

FIG. 42.—Terra-cotta spindle-whorl marked with swastikas. Troy, 4th city (Schliemann).

The Mykenæans and their island relatives obtained the Swastika either from the ancient Bronze-age people of Europe or else gave it to them, since it is very nearly as common as a decoration or symbol on the bronze swords, spear-heads, shields, and other metal work of these prehistoric people of the middle and north of Europe (also occurring in the pottery of the Swiss Lake dwellings), as it is in the islands and adjacent lands of the Eastern Mediterranean. The Swastika is also found abundantly on the early work of the Etruscans, but disappeared from general use in Italy, as it did from the rest of Europe, before historic times, although occasionally used (as in the decoration of the walls of a house at Pompeii). All over Germany, Scandinavia, France, and Britain it is found (Fig. 44) on objects of the Bronze period—sometimes on stone as well as on bronze utensils, ornaments, and weapons. A few objects of Anglo-Saxon age are ornamented with it—especially remarkable is a piece of pottery of that age from Norfolk (Fig. 45).

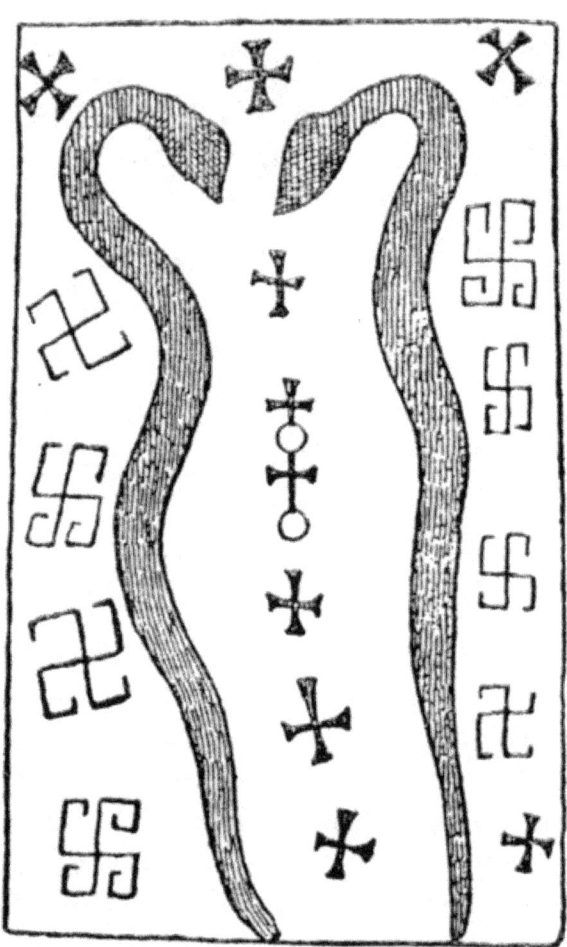

FIG. 43.—Ornament from an archaic (pre-Hellenic) Bœotian vase, showing several swastikas, Greek crosses and two serpents (from Goodyear).

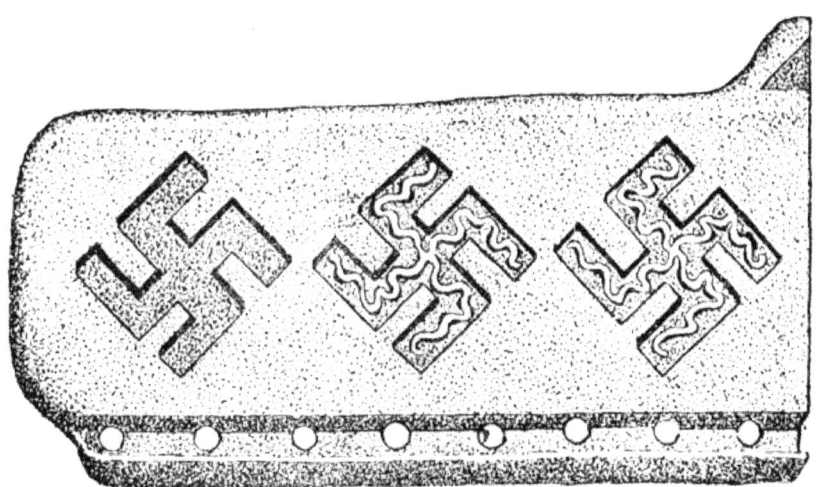

FIG. 43. (*bis*).—Swastika in bronze repoussé, from the necropolis of
Koban, Caucasus (after Chantre "Le Caucase"), about 500 B.C.

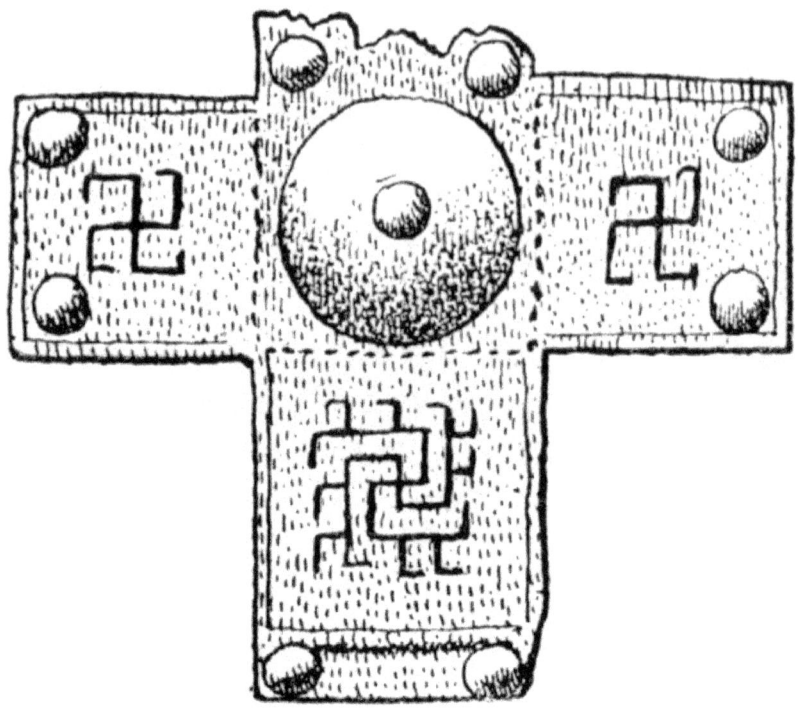

FIG. 44.—Silver-plated bronze horse-gear from Scandinavia, showing two swastikas, and below a complex elaboration of a swastika. (Bronze Age, about 1500 B.C.)

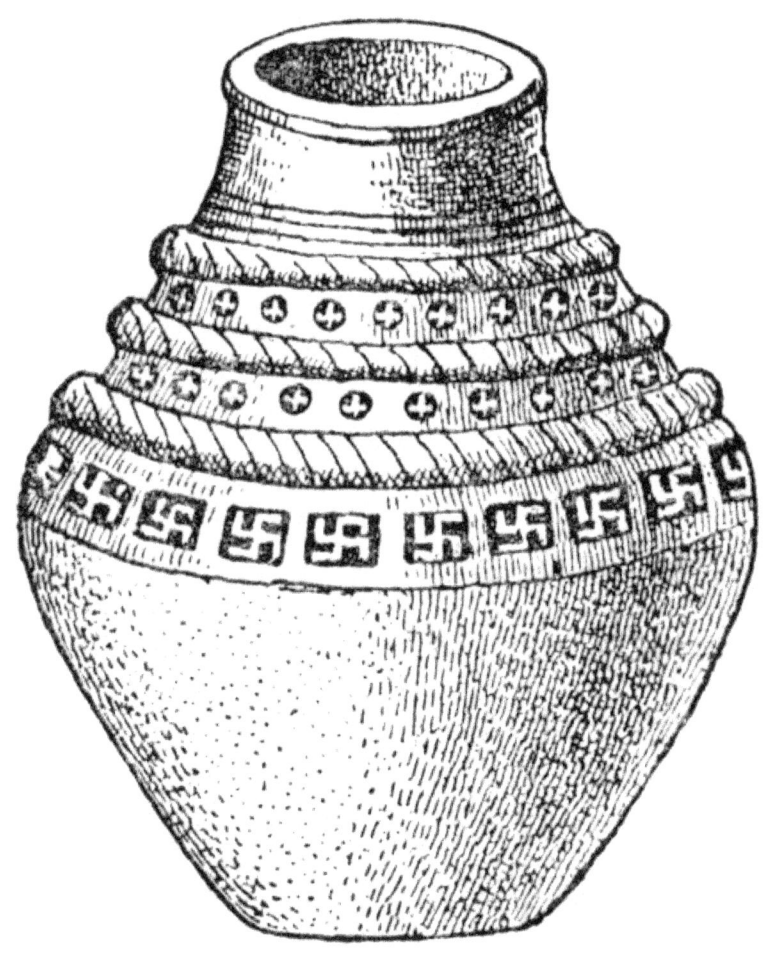

FIG. 45.—Anglo-Saxon urn from Shropham, Norfolk, ornamented by twenty small hand-made swastikas stamped into the clay. (British Museum.)

The history of the "Swastika" would be remarkable enough if it ended here with the disappearance of its use in Europe in prehistoric times and its continued use in the Far East and India. But the most curious fact about it is that we find it as a very common and favourite decoration and device among the native tribes in North America and Mexico, and exceptionally in Brazil. It is found in use among the Indians of Kansas and other tribes—as a device in pottery, in bead-work (Fig. 46), patch-work, quill-embroidery, and other decorative fabrics. The Indians called Sacs, Kickapoos, and Pottawottamies, who worship the sun (which is associated with the Swastika in China), call it by a native name signifying "the luck." It is also found as a decorative design in the most ancient remains of man in America, dating (so far as can be guessed) from a thousand years or more before Columbus (Fig. 47).

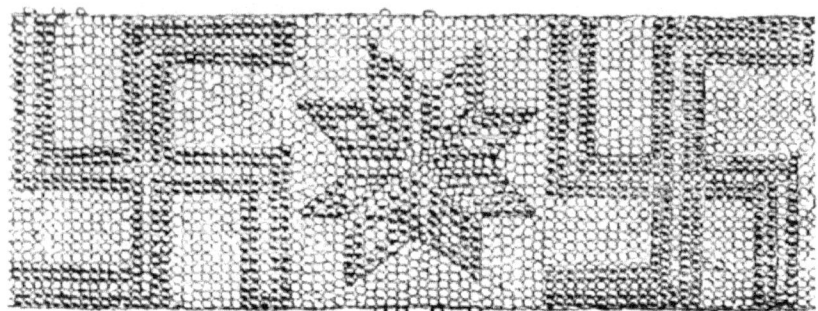

FIG. 46.—Piece of a ceremonial bead-worked garter, showing star and two swastikas made by the Sac Indians, Cook County, Kansas. (Modern.)

It is generally held that the Swastika must have been introduced into America in prehistoric times by early redskin immigrants from Asia. The question has been raised as to whether this introduction was before or after the worship of Buddha in Asia. It is only amongst Buddhists that the Swastika has a religious or sacred character. Elsewhere it seems to have been a mark or sign carrying "good luck." A representation of a sitting human figure incised on shell has been found in a prehistoric burial-mound in Tennessee, which has remarkable resemblance to the Asiatic statues of the Buddha. Shell ornaments have also been found here decorated with sharply-cut Swastikas, and in a mound in Ohio thin plates of copper were found cut into simple Swastika shapes like that of Fig. 38, four inches across. Modern Mexican Indians make brooches of gold and turquoise in the form of the Swastika, and it is a favourite device among the

Indians of neighbouring territory. Swastikas occur as decorations or lucky marks on the small terra-cotta "fig-leaf," which was worn by the women of some of the aboriginal tribes of Brazil, and have also been found on native pottery from the Paraguay River.

Some students of this subject have held the opinion that the "Swastika" has been invented independently at different times in different parts of the world. It is a fairly simple device, it is true; but the view which is accepted at present is that it has spread from one centre—probably European in the late Stone period—through the Mykenæans, across Asia, and so with early immigrants across the Pacific into the American continent.

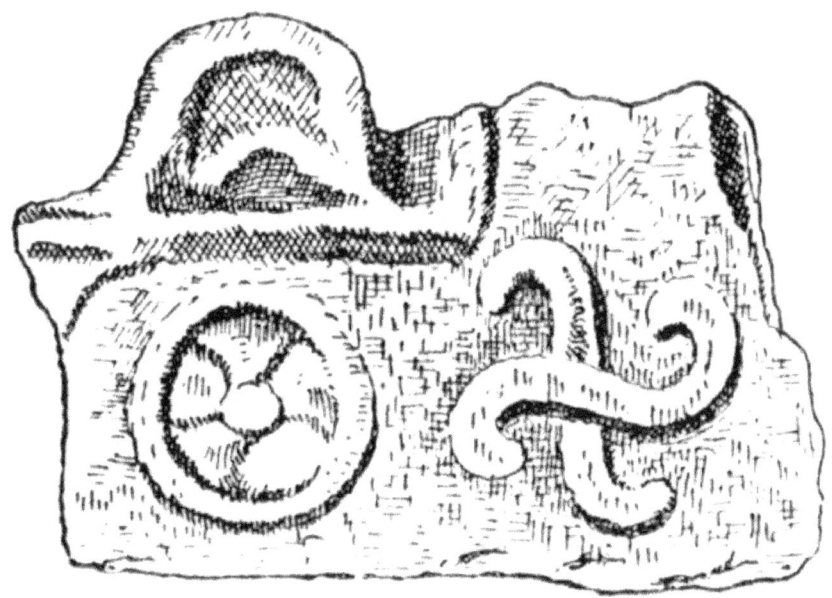

FIG. 47.—A stone slab from the ancient city of Mayapan (Yucatan, Central America), on which (right side) a curvilinear swastika is carved. (From the American Antiquarian Society, 1881.)

Apart from this problem, there is an interesting question as to how the device probably took its origin. The "Swastika" is sometimes called the "gammadion," because it may be regarded as four individuals of the Greek letter gamma (which has this shape [Greek: G]) joined at right angles to one another. The old English name for it, dating from Anglo-Saxon times, was fylfot—an old Norse word of doubtful meaning, which has no currency at the present day.

A method of making the Swastika by piling up sand or grain on a flat surface, actually in use at the present time in India, is shown in Fig. 48. The artist makes first of all a circle with a cross drawn within it (A). Then the circle is rubbed out or cut away at four corresponding points where the arms of the cross touch the circle, and so we get B. Then by the straightening of the curved pieces we get the correct rectangular Swastika, C. It is not probable that this is the way in which the Swastika was originally devised, though it is not possible to arrive at any certainty on the subject.

In these matters concerning the origin of simple ornamental patterns, designs, and symbols, we always have to deal with certain natural opposing tendencies on the part of the artist-draughtsman or designer, one or other of which may be variously called into prominence by the softness or hardness or other quality of the material he has to use, or by the individual fancy for elaboration or for simplification which exists in him. I will call four of these tendencies which concern us in regard to the Swastika: 1, the rectilinear as opposed to 2, the curvilinear, and 3, the grammatizing as opposed to 4, the naturalizing tendency, and will show what bearing they may have on the origin of the device known as the Swastika.

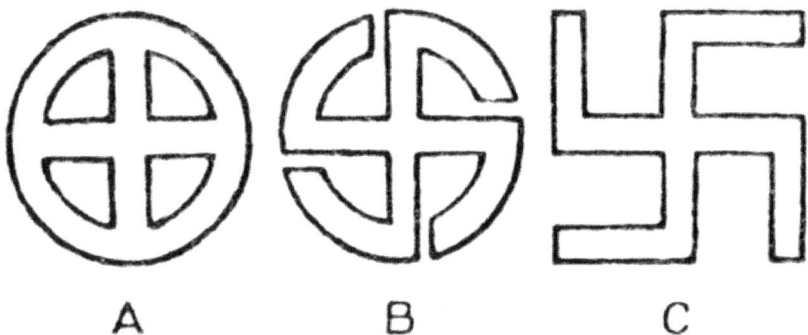

FIG. 48.—Diagram to show the derivation of the swastika from a Greek cross enclosed by a circle. In India the swastika is actually modelled in this way—in native ceremonies with rice-grain spread on the ground. The successive figures drawn above are produced by moving the rice with the hand.

CHAPTER XVIII

THE ORIGIN OF THE SWASTIKA

THE Swastika is, we have seen, a very early device or symbol in use among ancient races in Europe, Asia and America. Though it has been found on an ingot of metal in Ashanti it was of late foreign introduction there, and is not known in Africa, nor in Polynesia and Australia, nor among the Eskimos. How did it as a mere matter of shape and pattern come into existence? One might suppose that such combinations of lines as the simple cross and this modified cross, with the arms bent each half-way along its length to form a right angle, would be very natural things for a primitive man—or a child—to make when trying to produce some ornamental effect by tracing simple rectilinear and symmetrical figures. No doubt such a "playing with lines" is a common phase or stage of the human search for decorative design. It leads by gradual steps to very complex line-decoration in early pottery and woven work, which is sometimes called "geometrical design."

It is, however, the fact, and a very interesting one, that the tendency to make geometrical design is not so pronounced in the very earliest examples of human drawing and ornament known to us, as is the tendency to copy natural objects. And this would appear to be especially the case where the drawing is to be a symbol or significant badge. In the earliest art-work known to us—that of the cave-men of the late Pleistocene period in Western Europe (see Chaps. I., II. and III.)—the artists were busy with attempts (often wonderfully successful ones) to present the outlines of familiar animals (and sometimes plants) by incised carving on bone or painting on the rock walls of caves—preceded, it is true, by a period in which "all-round" sculpture in bone or stone or modelling in clay was the method employed. The extensive use of lines—concentric or parallel, like those on the finger-tips—as decoration of stone work is not known until the later or Neolithic period. [7] On one at least of the incised bone drawings of the Palæolithic cave-men two little diamond-shaped lozenges are engraved. They are seen in the cave-men's drawing of a stag figured on pp. 12, 13 of this book. These lozenges are supposed to be the "signature" of the artist, and, if so, are not only the first examples of a geometrical rectilinear figure as ornament, but the earliest examples known of the use of a badge or symbol as a means of identification.

When we compare the simpler decorative designs made use of by the less cultivated races of men, we find that there are certain distinct and opposed tendencies the predominance of which is of importance in helping us to explain the origin of the design. The tendency to make straight lines and rectilinear angles, which we may call the "rectilinear habit," is found in work executed on hard stone by a graving tool, and in work where square-cut stones are set together or flat pieces of wood or straw are interlaced, and in coarser kinds of weaving, bead-work, and basketwork. The opposite tendency is found in work executed with a brush and fluid paint on pottery or cloth, or even with a graver on soft clay or bone.

The contrast is well shown in the two renderings of one and the same "pattern," shown in A and B of Fig. 49. A is the rectilinear angular decorative design which is known as the "Greek key pattern," whilst the scroll below it is the "curvilinear" treatment of the same subject. The first takes its rectilinear character from a structure built up of hard blocklike pieces; the other is the flowing, easily moving line of a brush laying on paint, or of a style moving over clay or soft wax. The contrast is the same as that of the capital letters of the Roman alphabet, as used in print, with their equivalents in "copper-plate," cursive handwriting.

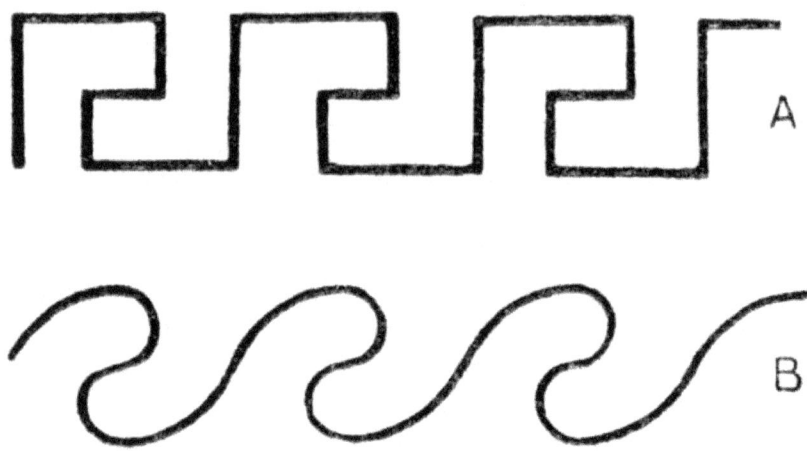

FIG. 49.—The Greek Key pattern in *A* rectangular, and *B* curvilinear or "current" form.

Another pair of tendencies opposed to each other which have much significance in the explanation of decorative design is the tendency to convert the simple lines of an original design into a drawing representing some animal or plant shape. At the end of the last chapter I distinguished this as the "naturalizing" tendency, contrasting it with the grammatizing or simplifying tendency. A good example of it is seen in Fig. 50. In A of that figure we see a circle divided into three cones by curved lines; this is a known design. It is called a "triskelion" (meaning a three-legged figure), or is more correctly termed "a three-branched scroll." The curves are converted into angles and straight lines in B, and then the stiff rectilinear "triskelion" is subsequently developed into three human legs, as shown in C, Fig. 50. It is naturalized. Were the change to proceed in the other way from the three human legs to the simple lines, we should have an example of the opposed tendency, namely, that of converting drawings of natural objects—by a degenerative or reducing process—to the simplest lines representative of them. This tendency, which we call "grammatizing" (from gramma, the Greek for a line), is far commoner in early art than the naturalizing tendency which sets in when the artist is exuberant, self-confident, and imaginative. We see a "naturalizing" tendency in the flamboyant and arabesque decorative work of the renascence, but it is also found among the happy Minoan, or Ægæan, island folk who decorated great pots and basins in Cyprus and Crete with forms suggested by birds, sea-creatures, and climbing plants, and worshipped the great mother Nature as Aphrodite, the sea-born goddess.

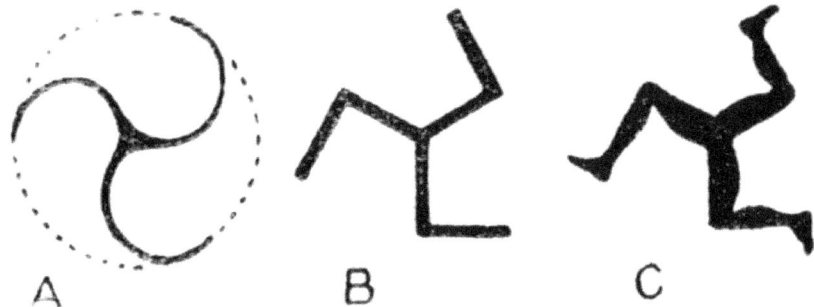

FIG. 50.—Diagrams of the "triskelion" or figure formed by the division of a circle into three equal bent cones as in *A*. *B* is the rectangular form derived from it. *C* is a "naturalized" form derived from it, namely, the three conjoined legs used as the badge of Sicily and of the Isle of Man.

The triangular island of Sicily (called also Trinacria) had in ancient times (even as far back as 300 B.C.) the conjoined three legs (shown in Fig. 50, C) as its badge or armorial emblem. An ancient Greek vase found at Girgenti has this badge painted on it. Ancient Lycia had a triskelion formed by three conjoined cocks' heads stamped on its coins. Though it has no direct connection with the Swastika, the introduction of the "three legs" as the armorial emblem of the Isle of Man is worth relating, as it is not known to most of those who are familiar with the device, with its motto, "Quocunque jeceris stabit" on the copper pence minted for that island up to as late a date as 1864, and current in Great Britain. King Alexander III of Scotland expelled the Norse Vikings from the Isle of Man in A.D. 1266, and substituted for their armorial emblem in the island, which was a ship under full sail, the three legs of Sicily. Frederick II, King of Sicily, married Isabella, the daughter of Henry III of England. Alexander III of Scotland married Margaret, another daughter of Henry, and Henry's son, Edmund the Hunchback, became King of Sicily, in succession to his brother-in-law Frederick. Alexander of Scotland was thus brother-in-law both of Frederick II and of Edmund, successive kings of Sicily. It was in this way that he was led, when he added the Isle of Man to his kingdom, to replace the former Norse emblem of the island by the picturesque and striking device of that other island—Sicily—with which he had so close a family connection.

The tendency for drawings of men and animals when used as decorative designs to degenerate, in the course of time and repetition, into more and more simple lines, to become more and more "grammatized" and simplified, till at last their origin is hardly recognizable, is both a very remarkable and a very usual thing. The process of degeneration, step by step, can often be traced, and curious remnants of important parts of the original drawing are found surviving in the final simplified design. The paddles and other carvings of some of the South Sea Islanders show very curious "degenerations" of this kind. A carved human head with open mouth becomes by repeated copying and simplification a mere crescent or hook, which is the vastly enlarged mouth of the original face. It alone survives, and is of enormous size, when all other features and detail have been abandoned. In some carvings of a face the tongue is shown projecting as an indication of defiance. In course of simplification in successive reproductions the face becomes a mere curved surface with a large pointed piece standing out from it; it is the tongue. That one significant thing—suggesting defiance—alone persists. The study of this process in human art covers a very wide field, including all races and all times. An excellent example is that given in Fig. 51. It shows the step by step "grammatizing" of a favourite decorative drawing—that of an alligator, as painted by the Chiriqui Indians of Panama on pottery. We start in Fig. 51, A, with an alligator, already considerably "schematized" or conventionalized. The Indians could do better than that, but it served for pottery decoration. The figures B, C, D show three stages of further "grammatizing" of the design (from different parts of the surface of a pot) till, in D, we get the alligator reduced to a yoke-like line and a dot!

FIG. 51.—Four stages in the simplification of a decorative design—the Alligator—as painted on pottery by the Chiriqui Indians. (Holmes.)

Familiar modern examples of this reduction of an animal figure to one or two lines, with mysterious-looking branches (representing limbs or horns), are seen in the scattered devices on the Turkey carpets so largely used at the present day. A comparison of various examples of such carpets of different age and locality reveals the true nature of these queer-looking patterns as representations of animals! Another familiar instance of the grammatizing of an animal form is that shown in Fig. 52, D, which is the common symbol in modern European art for a flying bird. Fig. 52 shows, however, some more important simplifications of animal form. The series marked E are a few examples from hundreds painted on the walls of caves in Cantabria (Spain) by prehistoric men. They start with a clearly recognizable figure of a man—many such, an inch or two high, occur on some parts of the cave-walls—and then we have all sorts of simplifications and deviations from the more naturalistic initial design, as shown by the rest of the series, ending in a T—a primitive symbol often arrived at by savage decorative artists in various parts of the world by reducing and grammatizing the human figure. The letters of many alphabets have been simplified in this way from original picture-like signs or pictographs.

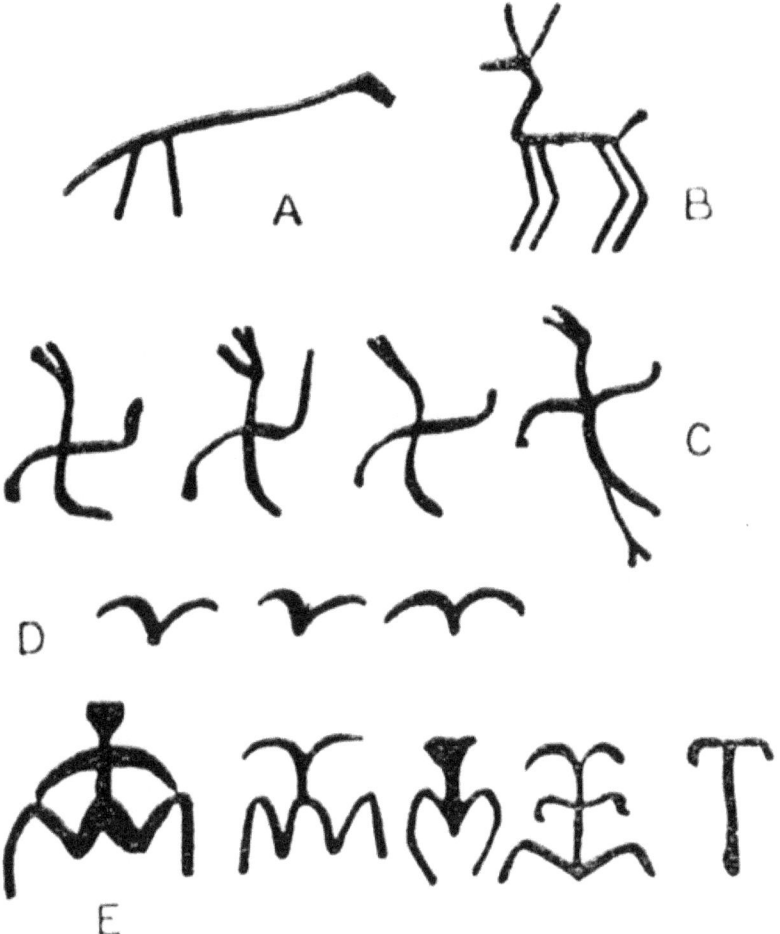

FIG. 52.—Simplification (grammatizing) of decorative design. *A*, a stork walking. *B*, a stag. *C*, a stork with wings spread for flying—resulting when fully "grammatized" in a curvilinear swastika. *A*, *B*, and *C*, from spindle-whorls found at Hissarlik. *D*, conventional representation of three flying birds. *E*, grammatized human figure from the walls of caverns in Cantabria.

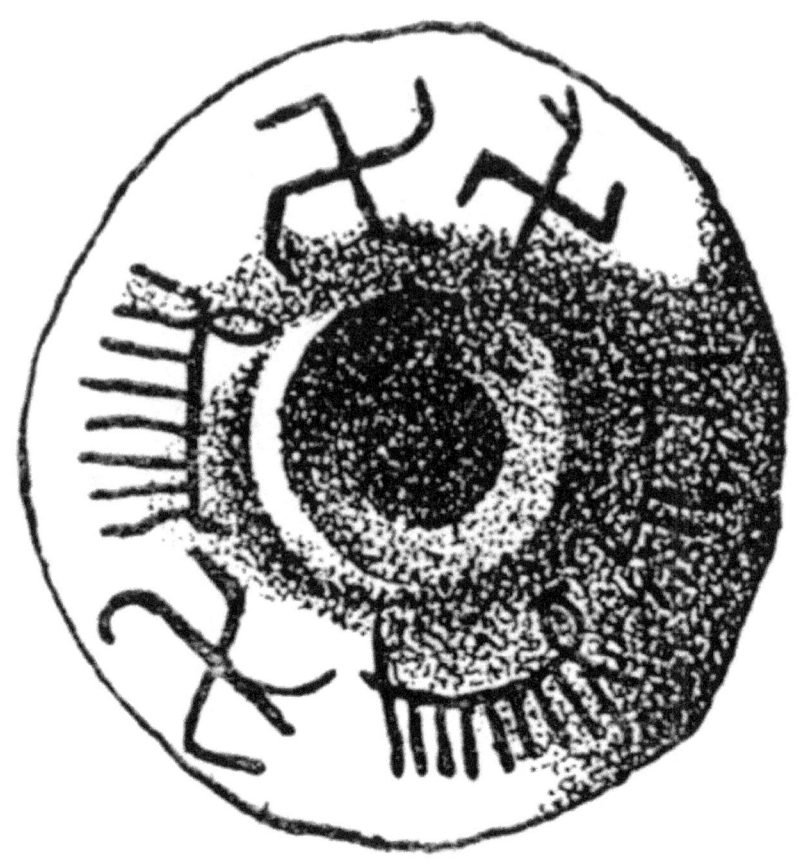

FIG. 53.—Spindle-whorl from Troy (fourth city), with three swastikas—
two resembling "stylized" storks (see Fig. 52, C). (Schliemann.)

The drawings lettered A, B and C in Fig. 52 represent
accurately figures scratched on the clay "spindle-whorls"
(before baking), so abundant in the remains of the ancient
cities on the hill of Hissarlik (Troy), found by Schliemann
(see Figs. 42 and 53). These heavy, bun-like spindle-
whorls have retained their use and shape since Neolithic
times (they are found in the Swiss lake-dwellings) to the
present day. Similar whorls were made of modern
porcelain, variously decorated, in France in the last century
and sold to the peasants for giving weight and rotatory
stability to the spindle used in spinning, and are still used
wherever the spindle survives, as among the Indians of

Central America. A "grammatized" profile representation of a stork (Fig. 52, A) is one of the designs on these Hissarlik spindle-whorls, and so is the linear representation of a stag (Fig. 52, B). And now we come back to the Swastika. The four figures in a row, marked C in Fig. 52, are a few of the representations of "flying" storks on these same spindle-whorls; one so marked is drawn in Fig. 53. They are of various degrees of simplification, and the last but one on the right hand side is identical with a Swastika! It must be carefully remembered that these clay spindle-whorls from Hissarlik are very commonly inscribed with undoubted well-shaped Swastikas, as shown in Fig. 42. The Swastika is quite a common and usual decorative lucky badge in the household art of that locality and age. Hence it is not surprising that M. Solomon Reinach, of Paris, has suggested that the Swastika may have originated thus—by the "stylizing" or "grammatizing" of a favourite and sacred bird—the stork. Once thus suggested and drawn in the simple Swastika shape the emblem (it would be supposed) became fixed, and made as rectilinear and simple as possible. Thenceforward it was accepted as an emblem of good luck, which has been transmitted throughout the ancient world of Europe, Asia and America. This theory has a plausible aspect, but I understand from M. Reinach that he no longer attaches importance to it. I do not know what theory, if any, of the origin of the Swastika now commends itself to him, nor whether he thinks it has originated independently in several times and places, or holds that it has one common origin. I am inclined to favour the theory that the Swastika has been started by the copying of the form of a natural object on the part of a primitive race of men, and that this form has lent itself to the invention of other badges and symbols besides that known as the Swastika. I will explain this in the next chapter.

FOOTNOTE:

[7] But spiral and leaf-like decorative designs engraved on bone (see Fig. 29, p. 54) are found in caves associated with other carvings made by cave-men of the Reindeer or late Palæolithic period.

CHAPTER XIX

THE TOMOYE AND THE SWASTIKA

FIG. 54.—The "Tomoye"—the Japanese badge of triumph.

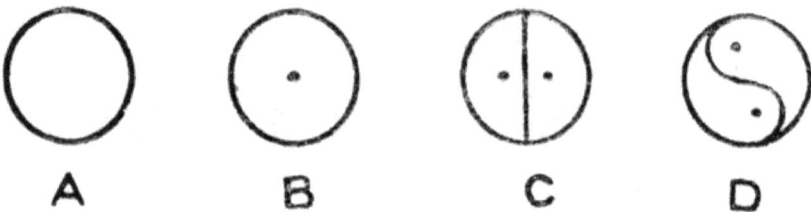

FIG. 55.—Symbols of the history of the universe used by the ancient Chinese philosopher Chu-Hsi. *A*, The original "void." *B*, The great monad. *C*, The monad divides into two, male and female. *D*, The halves in rotary movement, suggested by the S-like bending of the dividing line or diameter of the circle.

FIG. 56.—Diagrams to show the possible derivation of the swastika from the inscription of two S-like lines (or "ogees") within a circle so as to divide the circle into four bent cones. *B* and *C* are ogee and rectangular swastikas easily produced by modification of the encircled figure.

FIG. 54 represents a remarkable design which is a sort of national emblem, a universally accepted badge of triumph and honour in Japan, and is called "Tomoye"—meaning "triumph." The black and white portions are in that country painted respectively red and yellow. It is simply a circle divided into two equal cone-like figures by the inscription within it of a doubly-curved line like the letter S. Where and how did the Japanese get this badge? Who invented it, or from what natural object is it copied? A modified Tomoye with the cones dislocated is used as the national flag of Korea. A single one of these curious, tapering, one-sided cones is closely similar to the cone-like figures sometimes called "pines" which one sees on Indian shawls. The origin of these is sometimes said to be a copying of some fruit or vegetable growth, but is really not ascertained—and is possibly half of a Tomoye! A great circular altar-stone has been found in Central America, 5 ft. across, divided by a deep S-shaped groove into two equal one-sided cones (Fig. 59) like the Tomoye. The figure formed by an S within a circle is found in the writings of the ancient Chinese philosopher Chu-Hsi. He gives a series of symbols representing (according to him) the history of the universe. They are shown in Fig. 55, and are explained as follows. The empty circle A represents the original "void"—the boundary line is conventional. After untold æons the great monad appeared. It is represented by B. Then we get the division of the great monad (now called "Tai-I") into two, shown in C of our Fig. 55—singularly recalling the division of the nucleated cell or protoplasmic unit of animal and vegetable structure. The two halves, however, in this case represent the feminine called "Yin" and the masculine called "Yang." The last drawing, D of Fig. 55, shows the Yin and the Yang in rotatory motion. This is indicated by the S-like bending of the diameter, and the consequent formation of a figure like the Tomoye. By this motion the visible universe is supposed—by the philosopher Chu-Hsi—to be produced. The figure marked D is described as a "cosmological symbol." It does not help us to the origin of the figure showing the division of the circle as in the Tomoye, for it dates only from about the twelfth century of our era.

If we suppose the circle divided, as in the Tomoye, to be a very ancient badge or device, dating from prehistoric man, then it is probably derived from a natural object. And this object was probably a ground-down transverse section across a whelk-shell, for if one makes such a section just above the mouth of the shell at right angles to its length, one gets two adjacent chambers of the spirally-coiled shell separated by an S-like partition, the resulting figure given by the slice across the shell being that of the "tomoye," with its paired, one-sided, cone-like constituents. Shells are amongst the chief ornaments used by prehistoric and modern savage man. Large ones are ground down to make armlets. The perception of the spiral as a decorative line is almost certainly due to the handling and grinding-down of snail shells, and, indeed, we find spirals and reversed spiral scrolls engraved on bone by the Pleistocene cave-men (see Fig. 29).

FIG. 57.—Terra-cotta cone with a seven-armed sun-like figure engraving on it. Troy. (Schliemann.)

The Ægæan people of the Greek islands (of whom the Mykenæans are a part) copied a variety of forms of marine animals in their decorations of pottery, and, in fact, natural shapes were the basis of their decorative art. They simplified and "grammatized" their more nature-true designs into badges and symbols.

FIG. 58.—Scalloped Shell Disk, from a mound near Nashville, Tennessee, showing in the centre a tetraskelion with four curved arms, about four inches in diameter, made of polished shell. (Peabody Museum.)

We find in early work discovered in the ancient mounds of North America decorative circles (Fig. 58) in which two S-like lines at right angles to one another are inscribed as shown in Fig. 56, and we find also that these curved rays may be prolonged as a marvellous enveloping spiral coil or helix—especially in the painting of pottery. When the curved rays are many in number, as in Fig. 57, the design has been interpreted by some archæologists as symbolizing the sun, and it is important to remember that the Swastika itself was used in China as the pictograph of the sun. A single curved S-like line has been found cut on a great circular slab, an ancient altar-stone (Fig. 59) in Honduras (Copan)—so as to divide the circle as is done in the Japanese Tomoye. It is obvious that the exact geometric character of the S-like division is of great significance in these designs and requires careful study and explanation. I have briefly discussed this matter at the end of the chapter. In the common "ogee Swastika," Fig. 56, B, the more or less elaborately helicoid arms are merely careless flourishes of the painter's brush. The simple four-rayed figure, shown in Fig. 56, A, is often spoken of as a "tetraskelion," or four-legged scroll, and is associated with the three-legged figure or triskelion which I wrote of in the last chapter. If the curvilinear "tetraskelion" be angularized—that is to say, rectangles substituted for semicircles, we get the correct fully developed Swastika, Fig. 56, C. And if, abandoning the circle, the draughtsman rapidly drew with a brush or on soft clay lines like an S crossing one another at right angles, he produced what is common enough wherever the more formal rectangular Swastika is found, namely, the curvilinear or "ogee Swastika," Fig. 56, B.

It is not possible with our present knowledge to penetrate into the remote past and really ascertain the origin of the shape or device called a Swastika. But it is, I think, quite likely that in manipulating the "tomoye" symbol (whether copied from a section of shell or originating by more independent invention and "trying" of lines and curves and circles), very early man duplicated the symmetrical S by which he had divided a circle and produced the tetraskelion seen in Fig. 56, A. The conversion of this into the rectangular Swastika and into varieties of the ogee and menander (which I have not found space to describe) would be an easy and natural sequence.

FIG. 59.—An altar-stone of prehistoric age. The circular surface is cut into by a trough of S-shape, which divides it so as to resemble the Japanese "Tomoye." From Copan, Honduras.

At the same time, I have no conviction that this is the real origin of the Swastika, and await further evidence. The "flying-stork theory," which was put forward by Reinach, is very attractive. Birds as badges and "totems" are frequent among primitive mankind, and certain species are often regarded as sacred and bringing good luck. The stork is one of these. If the artists who marked the very ancient clay-pottery of Hissarlik with the Swastika and also with outlines of the flying stork, strongly resembling a Swastika, did not derive the Swastika from the stork, but had received it from some independent source, then it is probable that they purposely drew the flying stork, so as to make it resemble as much as possible a Swastika.

When we take account of the apparently arbitrary passage of human decorative design from the naturalistic to the linear, and from the linear to the naturalistic; from the curvilinear to the rectilinear, and from rectilinear to curvilinear; when we also reflect that some races and populations of men have been prone to seek for the forms of their decoration in the natural forms of plants and animals, whilst others have made use of mere mechanical patterns of parallel or interlacing lines, we must conclude that by the appeal to one or other of these various tendencies it is easy to invent a large variety of more or less plausible theories as to the origin of the Swastika. The truth of the matter can only be decided, if ever, by more direct and conclusive evidence than we at present possess. Nevertheless, it is a legitimate and fascinating thing to speculate on the origin of this wonderful world-pervading emblem coming to us from the mists of prehistoric ages, and to endeavour to arrive, if possible, at possible points of contact between it and other "devices" and "symbols," even though they may be of equally obscure birth. [8]

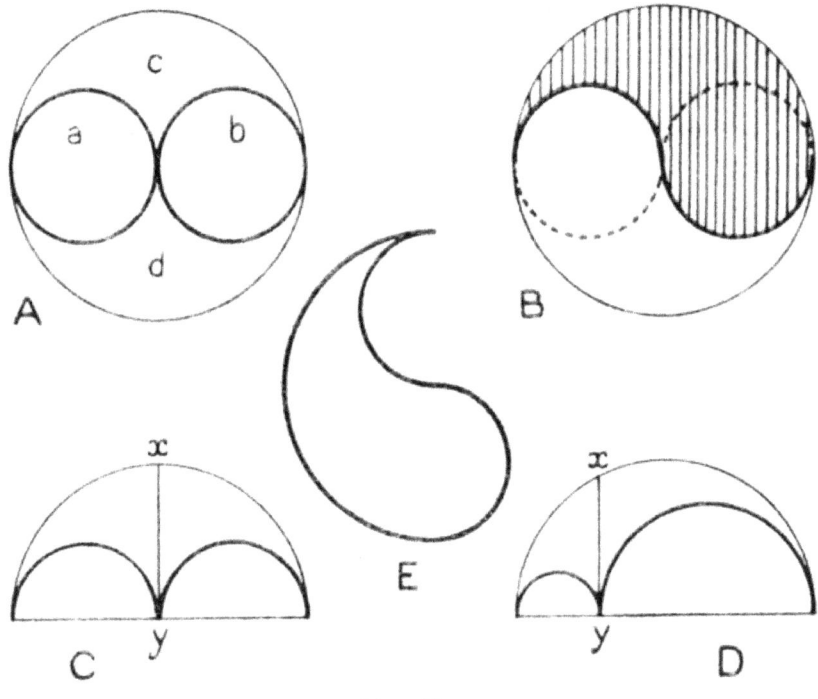

FIG. 60.

The accurate division of a circle into two equal comma-shaped areas of the special shape presented by the "Tomoye" of the Japanese (Fig. 54) and the rotating "Great Monad" of Chinese cosmogony (Fig. 55), is effected by describing within a given circle two circles each having its diameter equal to a radius of the enclosing circle. The two inscribed circles touch one another at the centre of the latter, but do not overlap. The area of the enclosing circle is thus divided into four areas, *a*, *b*, *c* and *d* (see Fig. 60, A). The areas *a*, *b* are the two inscribed circles. Each of the residual areas *c*, *d* is called (as Sir Thos. Heath, F.R.S., kindly informs me) an "arbelus" by ancient Greek geometricians—a name used for a rounded knife used by shoemakers. The comma-shaped bent cone or pine is formed by the fusion of one of the two small circles with one of the adjacent arbeli (Fig. 60, B). The figure so formed which to-day is loosely spoken of as a "bent cone," a "pine," or a "comma," has never, so far as I can ascertain, received a name in geometry, nor in the language of decorative design or pattern-making. Nor has the S-like line made by the two semicircles separating the contiguous "pines" or "commas" received any designation though vaguely indicated by the word "ogee." The comma-like areas might conveniently be called "streptocones," and their S-like boundary "a hemicyclic sigmoid." As shown in Fig. 56, by drawing a second hemicyclic sigmoid of the same dimensions at right angles to the first, the circle is divided into four smaller streptocones. By using sigmoids or half-sigmoids of a curvature of a different order from that of the hemicyclic one, but of a precisely defined nature, the circle may be divided into three, six, eight or more equal "streptocones" of graceful proportions, some of which have been used either in series as borders in metal work (for circular dishes and goblets) or as detached or grouped elements in pattern-designs (stone-work tracery, embroidery, woven and printed fabrics).

Apart from this development of the "streptocone" as an important feature in decorative work, it is not without interest in connection with the probable importance and significance of the Japanese double streptocone, as we may call the Tomoye, to note some of its geometrical features. Referring to the Fig. 60, it is obvious that each of the paired streptocones is equal in area to half the enclosing circle, also that each of the two inscribed circles (*a*, *b*) has an area of one-fourth of that of the enclosing circle—and that each arbelus (*c*, *d*) has also an area one-fourth that of the enclosing circle and is equal in area to each of the inscribed circles (*a*, *b*). Each of the two constituent "streptocones" is made up of a *complete* circle capped by an "arbelus" equal in area to it (namely, one-quarter of that of the big circle). It is obvious that the area of the arbelus formed in a semicircle by two enclosed semicircles which are contiguous and of equal base as in Fig. 60, is equal to that of a circle the diameter of which is the vertical line drawn from the apex of the arbelus to the arc of the semicircle (Fig 60). This is true whether the enclosed contiguous semicircles have chords of equal or unequal length (Fig. 60). This fact was known to the Greek geometricians, as I am informed by Sir Thos. Heath.

FOOTNOTE:

[8] I am indebted for the figures (not the diagrams) illustrating Chapters XVII., XVIII., XIX. to the report by Mr. Thomas Wilson on the Swastika—in the Smithsonian Reports, 1894. Those interested in this subject will find a vast store of information in that report.

CHAPTER XX

COAL

COAL is so much "a matter of course" in our daily life that most people are only now, when its supply is becoming precarious, anxious to know something of its nature and history. By the word "coal," or "coles," our ancestors understood what we now distinguish as "charcoal," prepared from wood by the "charcoal-burner," or "charbonnier," as the French call him. What we now call "coal" was known to them as "sea-coal," and, later, as "black" or "stone cole," to distinguish it from "brown coal," known nowadays as "lignite," though the name "stone coal" is locally applied in England to that very hard kind of black coal also called "anthracite," of which jet is only an extremely hard and dense variety found in small quantities in the oolitic strata of Whitby, Spain, and other localities.

It is on record that in the year 1306 a citizen of London was tried, condemned, and executed for burning "sea-coal." This severe treatment was held to be justified by the poisonous and otherwise injurious nature of the smoke produced by fires of sea-coal. I have not met with any records of the earliest digging for and trade in "sea-coal," but presumably it was obtained near the coast in the North of England and brought to London by ship—hence its name. The coal-trade of Newcastle began in the thirteenth century, but, owing to an Act of Parliament in the reign of Edward I forbidding the use of sea-coal in London, did not become important until the seventeenth century. It came very gradually into use, and we find that Evelyn (the diarist) in 1661 noted the withering and bad condition of rose-bushes and other plants in London gardens, which he attributed to the pestilential action of the smoke given off by the newly introduced "sea-coal" which was increasingly used as fuel in London houses. The sea-coal was not yet largely, if at all, used in the production of iron; and Evelyn as a forest-owner and lover of trees, has much to say about the necessity for attention to the cultivation of our forests in connection with the iron industry which then flourished in the Weald of Sussex; charcoal procured by the slow burning or roasting of wood being the fuel used in the smelting furnaces, whilst the ore was the orange-brown wealden sand. It was during the eighteenth century that what we now call simply "coal" came rapidly into use— not only for domestic heating, but for furnaces of all kinds employed in industrial enterprise, and, at a later date, for the earlier and later forms of steam-engines. The smoke of the new coal was everywhere regarded as a terrible nuisance, and a source of injury to both animal and vegetable life. The poisonous action of coal-smoke is not due to the finely divided black particles of carbon of which it largely consists, but to the sulphuric acid derived from the small quantities of sulphur present in coal. It is calculated that more than sixteen million tons of coal are annually used in London alone for heating purposes, and that 480,000 tons of black carbon powder are discharged over London by its chimneys every year, together with very nearly the same weight of poisonous sulphuric acid!

What, then, is this "sea-coal" or "coal" of our modern life? We all know its black, glistening appearance, and more or less friable character. Its nature and origin are best conveyed by the statement that it is very ancient "peat," compressed and naturally changed by chemical action and retaining little or no trace of its original structure. Peat, as we know it from the low land of English and French river valleys and the bogs of Scotland and Ireland, is formed by the annual growth and death of "mosses" of several kinds and of other accompanying vegetation. It retains the woody forms of the vegetable growths which constitute it, and they are often but loosely adherent to one another. Peat may be merely a growth of the past five years, but is sometimes many thousand years old. Older than peat, and more caked and compressed, is lignite, or brown coal, which occurs on the Continent of Europe, also in South Devon and elsewhere, in geological strata newer than those which yield our black coal. Then we have the most important class of black coals which are known as "bituminous coals," because they soften when heated and form hydrocarbons of both viscid and gaseous nature. They are used for domestic purposes, and wherever flame is desired. They are, in fact, the "lumps of coal" familiar in our scuttles. The "bituminous coal" with the greatest amount of hydrogen in it is the cannel or candle coal, so called from its bright flame when burning. This kind is especially valuable for gas-making, and of smaller value as fuel. The term "anthracite" is reserved for a hard, stone-like coal which is very nearly pure carbon (ninety per cent). This class of coal burns with a very small amount of flame, gives intense heat, and no smoke. It is used in drying malt and hops.

Like all woody matter, that from which peat is formed consists of a combination of the elements carbon, hydrogen and oxygen; and these remain in somewhat changed chemical union in the brown coal, bituminous coal, and anthracite. The carbon and a varying and small proportion of the original hydrogen of the woody peat, are the important elements in coal; and we may well ask how they come to be produced as a black or dark brown mass from dead vegetable growths which are often bleached and colourless. It is true that vegetable refuse does not necessarily blacken when left to itself. We know that by roasting or charring wood (or animals' flesh or bone) we can drive off the elements oxygen and hydrogen and nitrogen (if there), and obtain a black mass of carbon (so-called charcoal). That blackness is the actual true tint of carbon. The dead weeds and leaves at the bottom of a stagnant pond break down and form a pitch-black mud. They would not, and do not, go black if exposed to the oxygen of the atmosphere; but at the bottom of a stagnant pond or in a refuse heap they are excluded from the air, and a microbe—a bacterium which has been carefully studied, and is of a kind which can only flourish in the absence of free oxygen—attacks the dead weeds, producing by change of their substance marsh-gas and black carbon, the black mud emitting bubbles of gas which one may stir up with a pole in such a pond. This chemical attack by anaërobic bacteria goes on in the deeper layers of all marshes and stagnant pools, remote from the oxygen of the air; and it is fairly certain that the black coal which we find in strata of great geological age was so produced by the action of special kinds of bacteria upon peat-like masses of vegetable refuse. Indeed, by studying microscopic sections of coal, numerous forms of bacteria have been recognized which might be capable of effecting such chemical changes. On the other hand, we must remember that it is not possible to conclude by form alone as to what subtle chemical work a bacterium or bacillus or micro-coccus may be, or may have been, carrying on. The peat-like deposits which became carbonized and so formed the "coal" were probably masses of algæ, mosses and soft aquatic plants, which were brought down and accumulated in swampy, forest-covered ground about the mouths of rivers, the deposit being covered in owing to rapid oscillations of level by beds of sand or clay, followed by new growth and deposit.

Our British coal and a good deal of foreign coal is found in certain stratified rocks of the earth's crust known as "the

Carboniferous System," about 12,000 ft. thick, consisting chiefly of very dense limestone. The "seams," or stratified beds of coal, occur in sandy rock known as the "Coal Measures," and vary in thickness from a mere film to 40 ft. Above the Carboniferous System are later deposits, some 14,000 ft. in thickness—the Permian, Triassic, Jurassic, Cretaceous, and Tertiary strata. Below them we find stratified deposits containing fossilized remains of plants and animals, to a depth of another 40,000 ft.: they are the Devonian, Silurian, and Cambrian "systems" or series of strata. Coal of a workable nature is found in many parts of the world in the beds or strata of later age than our Coal Measures—namely, those of Jurassic, Cretaceous, and Tertiary age.

Coal is so valuable and used in such vast quantities by modern man that, though procured at first from beds lying at or near the surface, it has been found remunerative to mine far into the depths of the earth's surface, where its existence is ascertained, in order to procure it. A depth of 4000 ft. is apparently the limit set to such mining by the increase of temperature in mines which penetrate to that extent below the surface. In 1905 the annual output of British coal-mines was in round numbers 230,000,000 tons. It is certain that there is a limit to this production, but not possible to calculate what that limit may be, owing to the uncertainty as to the future working of coal-fields as yet unexplored.

Such questions have been, and are being, considered by experts on behalf of the Government. A matter of interest of another kind is that in and associated with the coal seams of our Coal Measures, fossilized remains of peculiar fern-like trees, ferns, and other strange plants, and of very peculiar, extinct newt-like animals (as large as crocodiles) are found in great variety. The notion that the toads occasionally found embedded in the black mud of a coal-yard or even in a fractured lump of coal are survivals from the time—many millions of years past—when the plants and animals of the Coal Measure swamps were living, is a baseless fancy. The toads so found are of the kind or species now living on the earth—totally different from those whose bones occur in the Coal Measures, and the presence of such modern toads embedded in black slime, in coal-heaps in store-yards, or even in coal-scuttles, is only what may be expected to occur and does occur in damp quarries and other places where these familiar little beasts love to hide.

CHAPTER XXI

BORING FOR OIL

CLOSELY similar to coal in chemical matter—that is to say, consisting chiefly of definite chemical compounds, called hydrocarbons, built up of only two elements, carbon and hydrogen, and of no other—is a very remarkable class of mineral substances known to the ancients as "bitumen." In its widest sense, it includes "natural gas," the variously mixed liquids called "petroleum" and the solid "asphalts." In ancient times the more fluid kinds of petroleum issuing from the ground in South Russia and Persia were called "naphtha," and that name is still applied to the more volatile hydrocarbons obtained by the distillation of such substances as coal-tar (the residue of the extraction by heat of commercial gas from coal), bituminous shale, petroleum, wood and some other bodies which owe their existence to the activity either of living or of long-extinct and "fossilized" plants and animals.

The bitumens, together with coal, present in their natural state a very large variety of inflammable constituents—gaseous, liquid, and solid hydrocarbons; but, when "distilled" at various temperatures and under conditions determined by the manufacturing chemist, they yield a still larger series of pure separable bodies, which have been minutely studied and classified according to their chemical constitution. They are produced in great chemical factories in large quantities for use in the most diverse ways invented by human ingenuity. Thus natural gas—superseded by distilled coal-gas—has served for fuel and for illumination: refined petroleum serves not only for those uses in general, but as the special source of power in the engines of motor-cars and aeroplanes. A wonderful solid crystalline wax-like substance, paraffin, as white as snow, is distilled in enormous quantities (nearly three million tons a year) from "bituminous shale" or "oil-shale" in this country alone. It can be obtained in soft (vaseline) and liquid forms, and in fact the "paraffin series" recognized by chemists starts from the gas "methane," or marsh-gas, and comprises some thirty kinds, leading from gas to volatile liquids, thence to viscid liquids, to butter-like solids, and up to hard crystalline substances which melt only at the temperature of boiling water. Endless chemical manufacturing industries—*e.g.*, those of dye-stuffs and explosives—depend upon the chemical treatment of these paraffins and of various bodies obtained as secondary products in their preparation. Benzine and aniline are chiefly obtained from coal-tar. The oils and waxes of quasi-mineral origin have a great advantage over vegetable and animal oils in many uses, since they are not liable to become "rancid"; that is to say, to decompose owing to the action on them of bacteria. A marked difference between the paraffins (often distinguished, together with the "olefines," as "mineral" oils) and the oils and fats found in living plants and animals is that they do not "saponify"; that is to say, they do not form those combinations with alkalis and other bases which are called "soaps," nor can they serve as food to man or any other animal. They are not acted on by the digestive juices.

From ancient times natural deposits or outpourings of "bitumens" have been known and used by mankind. The Assyrians and other early peoples of the East used "asphalt" (translated by the word "slime" in the English version of the Bible) in place of calcareous mortar in building; and to this day it is used largely in this country as a "damp-course" in walls built of brick. Great deposits of asphalt are found in Central America and some of the West Indian islands, and "quarried" for commercial purposes.

The great pitch-lake of Trinidad yields an abundant supply. In the Val de Travers, in the Canton of Neuchatel (Switzerland), a rich deposit is worked which, mixed with earthy material, forms a road-making concrete, largely used in London and other cities, and also for main roads in country districts. The ancient Egyptians used asphalt for embalming the dead. But the ancients also knew natural springs of liquid bitumen—that which nowadays we call petroleum—some of them freely flowing like water, which would take fire and burn for long periods, and were described as fountains of "burning water." We find, as we pass from the Middle Ages to the days of geographical exploration, records of such springs of inflammable oil and of natural inflammable gas in all parts of the world—Japan, China, Burma, Persia, Galicia, Italy (Salsomaggiore), Central and North America, and of not a few in these islands—for instance, in Shropshire, Derbyshire, Sussex, Kimmeridge and various sites in the southern counties. The oil was, until the middle of the last century, valued chiefly as a medicinal application, and "Seneca oil" and "American medicinal oil" were largely sold and used as an embrocation in the United States.

We owe the introduction of the name "petroleum" to Professor Silliman, who in 1855 reported upon the "rock oil or petroleum" of Venango County, Pennsylvania. The first attempt as a commercial enterprise to obtain rock-oil or petroleum by *boring* into the strata in which there was local evidence of its existence in greater or less quantity, was made in 1854 by the Pennsylvania Rock Oil Company. After some unsuccessful attempts, when the drilling had been carried to a depth of 69 ft. the tools suddenly dropped into a subterranean cavity, and on the following day the well was found to have "struck oil," and twenty-five barrels a day were yielded by that well for some time. From here the industry spread over the States and Canada, and in 1908 the year's yield was 45,000,000 barrels.

Since 1870 the industry has spread all over the globe—Russia, Galicia, Rumania, Java, Borneo and Burma being prominent sources of the oil supply of the world. The raw petroleum of different localities differs in each case in the amount of solid paraffins and olefines dissolved in the liquid paraffins. Other substances also are dissolved in it in variable amount—such as benzene, acetylene, camphene and naphthalene. The fact that the oil, when reached by a boring, is often found to be under a considerable pressure, so that it rises and flows from the surface of the well, or even may shoot up as a great fountain, is an important feature in the oil-seeking industry, though the supply depends largely on pumping and not necessarily on natural flow. The borings when made, act like Artesian wells, and sometimes are carried to a great depth. Those in Pennsylvania vary in depth from 300 ft. to 3700 ft., according to the distance below the surface at which the oil-bearing strata (usually a sandstone) is situate. As in the case of an Artesian well, the boring is in the first instance an exploration subject to uncertainty as to "striking" the desired liquid, but the uncertainty is greater in the case of the search for oil than in that for water. The water-well is also far less likely to "give out" when once flowing than is that bored for oil, which, even if at first successful, may be soon exhausted owing to the small area of the oil-bearing strata tapped. A cause of the high pressure in many oil-wells is the gas which accompanies the oil. The pressure may amount to as much as 1000 lb. to the square inch. In the Northern Caucasus spouting wells caused by the high pressure of gas in the boring are frequent. A famous fountain-well in that region, which began to flow in August 1895, threw up 4-1/2 million gallons a day, gradually diminishing during fifteen months until it became exhausted. At first, when boring was introduced, such outbursts led to an enormous loss of the oil, for there was not sufficient means of storing or transporting it. Ordinary cartage in barrels was the earlier method; then followed tanks on railway trains and canal boats; and this has been supplemented by the use of pipes along which the oil is pumped from the well to the refinery. In Pennsylvania there are said to be no less than 25,000 miles of such pipes in use for the distribution of petroleum.

It will be obvious from what is here stated that the attempt to discover an oil-supply in Derbyshire must not be regarded, at present, as more than a praiseworthy and interesting enterprise. There is no room for doubt that the best expert opinion has been brought to bear on the matter. A small quantity of petroleum has already been raised; but whether the flow will be sufficient to cover the expenses of the boring, and how long the flow may last, or how much it may amount to, are matters quite impossible to foretell. In any case, it is in the highest degree improbable that such an abundance of oil will be obtained as to count much, if at all, in the world's production of petroleum. It must also be remembered that products similar to those yielded by petroleum are already extracted in quantity as a remunerative industry by the distillation of oil-shales in various parts of the United Kingdom; and that there are oil-shales in this country still unworked. So that we need not be in despair if we do not tap an oil-spring of any importance close to hand. The world's supply is still open to British enterprise. Another reflection of some importance is that these world-wide sources of rock-oil or petroleum are likely to be exhausted by exploitation much sooner than are the coal-fields of the world. We cannot rely on their long duration.

CHAPTER XXII

THE STORY OF LIME-JUICE AND SCURVY

FROM mediæval times onward a serious constitutional disease—a morbid condition of the blood and tissues—has been known by the name "scurvy," and the word "scorbutic" has been coined from it. It is to-day practically unknown in the ordinary conditions of civilized life, but formerly was common, and the cause of disablement and of frightful mortality in ships' crews, beleaguered cities, armies on campaign, and war-stricken regions. It begins with a certain failure of strength. Breathlessness, exhaustion, and mental depression follow. The face looks haggard, sallow, and dusky. After some weeks the exhaustion becomes extreme; the gums are livid, ulcerated, and bleeding; the teeth loosen and drop out; purple spots appear on the skin; ulcers break out on the limbs; effusions of blood-stained fluid take place in the great cavities of the body; profound exhaustion and coma follow; and death results from disorganization of the lungs, kidneys, or digestive tract. It was recognized in early times that the disease was dependent on the character of the food of those attacked by it; and not the least of the horrors accompanying it was the terror caused by the well-founded conviction that the appearance of a single case in a ship's crew or other specially circumscribed community was an unfailing index, and meant that all were likely within a few days—owing to the enforced identity of their food and conditions of life—to develop the disease. Often, in past centuries, a half or two-thirds of a ship's company have been carried off by it before a port could be reached and healthy food and conditions of life obtained. At the present moment in view of the actual condition of Europe, it is a fact of very grave importance that scurvy is known to break out and cause a terrible mortality among civil communities in time of scarcity—especially in prisons, workhouses, and other public institutions, which are the first to suffer deprivations when food is scarce.

Three hundred years ago it was held that fresh vegetables and fruit-juices were both a cure for and a preventive of scurvy, or "anti-scorbutic." But the fact was not appreciated by Army and Admiralty officials that *dried* vegetables, even of kinds which were held to be especially "anti-scorbutic," would not serve in place of *fresh* ones. In 1720, *dried* "anti-scorbutic" herbs were supplied to the Austrian Army when suffering from scurvy; but they were of no avail, and thousands of the soldiers perished from the disease. A few years later, the British Lords of the Admiralty (actuated by a spirit of blundering parsimony) proposed to supply the Navy with dried spinach, although it was well known that dried vegetables were useless against scurvy. In the American Civil War, 1861-1865, in spite of this knowledge, large rations of dried vegetables were supplied to the armies, and failed to prevent outbreaks of scurvy. Even at the present day so little attention has been given of late years to the subject, that many ignorant officials, upon whose action the life of thousands depends, regard dried vegetables as equivalent in value to fresh!

A great advance was made in the second half of the eighteenth century, when the British Admiralty became convinced by the repeated experience of its officers that "lime-juice" *is* a specific remedy and preventive for scurvy, and, in spite of the great expense and difficulties entailed, adopted its use officially. In those days of sailing-ships, long voyages (such as those of Captain Cook) were safely carried through without serious outbreak of scurvy so long as a ration of so-called "lime-juice" (about one ounce) was swallowed each day by each sailor. But it was not until the beginning of the nineteenth century that the disease was practically eliminated from the Navy by the introduction (after many foolish delays) of a general issue of what was called "lime-juice."

The complete control and elimination of scurvy by the use of so-called "lime-juice" sufficed to carry us on until the introduction of steam navigation, when it became superfluous owing to the fact that long absence from land, where fresh food could be obtained, ceased to be usual. Moreover, after a mutiny on the part of our defrauded sailors, better food and greater variety of it was secured for them, and the profits of murderous contractors were stopped.

The history of outbreaks of scurvy for the last century is practically confined to the experiences of Arctic Expeditions and the campaigning of troops in remote or devastated regions. So little had scurvy been investigated, or any serious study made of the nature of the remedial and preventive action of lime-juice, that up to the year 1914 it was regarded as a matter of course that the acid, the citric acid, of lime-juice was what gave to it its virtue, and samples of lime-juice supplied by contractors were tested solely as to the percentage of that acid present. Eminent medical authorities proposed to use crystals of citric acid in place of the juice; others declared that vinegar would do just as well; others, in spite of the overwhelming record as to the value of lime-juice, held that scurvy was due *not* to the absence of a food constituent—supplied by fresh vegetables and fruit-juice—but to a peculiar poison present in the salted and dried meat served out as rations; others again, without any study of the disease, have expressed the opinion that it is due to a bacterial micro-organism.

A blow to the easy-going belief of the Admiralty that they had mastered and made an end of scurvy was struck when scurvy broke out (60 cases among 122 men) in the expedition to the North Pole which sailed in May 1875 in the *Alert* and the *Discovery*, under the command of Sir George Nares. The expedition had to return prematurely after seventeen months' absence, and a committee was appointed to inquire into the cause of the outbreak. The stores of food and of lime-juice were shown to have been ample; and the action of the leader in equipping his sledging parties was in accordance with the judgment and experience of successful explorers who gave evidence. The cause of the outbreak remained a mystery. The firm belief in the anti-scorbutic powers of "lime-juice" was shaken, and this unfavourable opinion of its value has been confirmed by medical officers who, during the recent war, have been confronted by outbreaks of scurvy. These outbreaks occurred among troops who, in military circumstances which rendered an adequate supply of fresh meat and vegetables impossible, were supplied with lime-juice prepared from the West Indian "sour-lime."

Under these circumstances, an experimental study of scurvy has been carried out during the last four years by a group of workers at the Lister Institute, together with a historical inquiry as to the use of lime-juice. The reports of these investigators have very great practical value and far-reaching interest, as showing what disastrous results may arise from inaccurate use of a word, and the neglect to ascertain the exact nature of the material thing upon which the issue between life and death may depend.

Here let me say that the staff of the Lister Institute for medical research has done work in its laboratories in Chelsea Gardens of the very greatest national importance during the war. It was founded by public subscription, and has now an endowment of some £10,000 a year. Sir David Bruce, the chairman of its Council, gives in the Report of the Governing Body for 1919 a very striking summary of the work done in the laboratories and by the staff of the Institute. The successful investigation of trench fever and of tetanus, of the destruction of lice, and of the effects of cold storage on food, besides the study of scurvy and other diseases due to deficiency of what are now called "*accessory food factors*," are, we learn, the chief matters in which the Lister Institute was engaged in the year 1918-19. Besides this, however, at its farm at Elstree it has prepared and supplied to the War Office, the Admiralty, the Overseas Forces, and the Local Government Board more than a million doses of anti-toxins (diphtheria and tetanus), bacterial vaccines (cholera, plague, influenza), and other similar curative fluids—requiring for their safe production the highest skill and most complete knowledge of recent discovery. And this is only a sample of what the Lister Institute has been doing for many consecutive years.

Now we return to the investigation of scurvy. Within the last ten years the fact has been established (which was more or less guessed and acted upon by medical men of past days) that, in order to maintain health, the diet of man and of many animals must contain not merely the necessary quantities of meat or cheese-like bodies, of fat and starch and sugar, but also minute quantities of accessory food-factors which it is convenient to term "vitamines." The name serves (though its etymology is unsatisfactory) to indicate certain "proteids" or highly complex nitrogenous compounds which are only to be obtained from fresh and uncooked or slightly heated vegetables and from some foods of animal origin. These "vitamines" are destroyed by heat and by desiccation. They have not yet been isolated though in some cases extracted in a nearly pure state. Their presence or absence is demonstrated by careful experiments in feeding animals, such as guinea-pigs, with weighed quantities of different foods. The "vitamine" is often found to be present only in one part of a seed or fruit or special kind of fat liable to be rejected in food preparation. An important fact is that it may not amount to as much as one-ten-thousandth of the weight of the food in which it occurs; and the part containing it may be overlooked and rejected, or its value destroyed by heat or by desiccation. A committee on these "accessory food-factors" is carrying on experiments at the Lister Institute. Dr. F. G. Hopkins, F.R.S., who first discovered the importance of one of these factors in feeding young rats, is the chairman, and Dr. Harriette Chick is the secretary. Three kinds of these vitamines, or accessory food-factors, have up to this date been recognized. The first is the anti-neuritic or anti-beri-beri vitamine. Its principal sources are the seeds of plants and the eggs of animals—yeast-cells are a rich source of it. Where "polished rice," as in the Far East, is the staple article of diet, to the almost entire exclusion of other food-stuffs, lassitude and severe pains like those of rheumatism set in, and a whole colony or shipload of Chinese "coolies" may be disabled. The disease is called beri-beri, and it can be cured by administering that part of the rice-grain (the skin and germ) which is removed by "polishing," and unfortunately is just that part which contains the needful vitamine. It exists in very minute quantity, amounting to only one part in ten thousand by weight of rice-grain. The second "vitamine" recognized is the anti-rachitic factor (studied by Hopkins), which tends to promote growth and prevent "rickets" in young animals. Certain fats of animal origin (milk) and green leaves contain it in minute quantity, and are necessary for the life of young animals and for the health of adults.

The third vitamine recognized is the anti-scorbutic, the factor which prevents scurvy. It is found in fresh vegetable tissues, and to a less extent in fresh animal tissues. Its richest sources are cabbage, swedes, turnips, lettuce, water-cress, and such fruits as lemons, oranges, raspberries, and tomatoes; other vegetables have a less value. Fresh milk and meat possess a definite but low anti-scorbutic value. This vitamine (I am quoting the report of the Committee, which has been issued to our military, naval, and medical administrators and famine-relief-workers throughout the world) *suffers destruction* when the fresh food-stuffs containing it are subjected to *heat*, or *drying*, as methods of preservation. It is habitually destroyed and wasted by stewing fresh vegetables with meat for two or three hours. All dry food-stuffs, such as cereals, pulses, dried vegetables and dried milk, are deficient in anti-scorbutic properties; so also are *tinned vegetables* and *tinned meat*—hence the disgust to which they soon give rise!

The explanation of the mystery about lime-juice (which a hundred years ago was used with absolute success to prevent scurvy, and in 1875 was a dead failure) is shown by the workers at the Lister Institute to be this—namely, "lime" and "lemon" are in origin the same word, and have become applied in ways unrecognized by the Admiralty and their medical advisers in various parts of the world to which the citron, the lemon, the sweet-lime and the sour-lime—all varieties of one species, *Citrus medica* of Linnæus—have been carried from their original home of origin, the south-east of Asia. The original effective and valuable "*lime*-juice" of the eighteenth century was *lemon*-juice, carefully prepared from lemons in Sicily and Italy, and from 1804 to 1860 in Malta. When the demand for it increased in the nineteenth century, it was adulterated and made up from poor fruit, as the commercial enterprise of contractors and the fatuous incapacity of the naval authorities progressed hand in hand. And then, in the early fifties, the West Indian growers of the small sour-lime (*Citrus medica var. acida*) in Montserrat got the naval contracts, the honest intention of Sir William Burnett, the chief medical officer of the Navy, being to establish a permanent and first-rate supply. Strangely enough, the naval "lime-juice" now really was *lime*-juice and no longer *lemon*-juice. By a natural but fatal misconception, the medical value of the juice, whether of lemon or of lime, was by all authorities attributed to the citric acid present; and the only tests applied to it were chemical ones, and not therapeutic. The Lister Institute Committee have shown by therapeutic experiment—the feeding of guinea-pigs, in which scurvy can be produced and cured at will—that *the anti-scorbutic vitamine remains active and unimpaired in lemon-juice from which all the citric acid has been extracted*. And, further, that the juice of the West Indian sour-lime (*Citrus medica acida*), although very rich in citric acid, *contains only one-fourth the anti-scorbutic vitamine* which the same quantity of the juice of the true lemon (*Citrus medica limonum*) contains. This has been most carefully established by prolonged series of feeding experiments. It explains the failure of the *lime*-juice in Sir George Nares' Polar Expedition, and restores the confidence in *lemon*-juice based on the unanimous testimony of the early records of its use.

Whilst lemon-juice is thus justified, Dr. Harriette Chick has made a discovery which will go far to remove it from supremacy. She finds that an anti-scorbutic food can be prepared, when fresh vegetables or fruit are scarce, by moistening any available seeds (wheat, barley, rye, peas, beans, lentils) and allowing them to germinate. This sprouted material possesses an anti-scorbutic value equal to that of many fresh vegetables; the unsprouted seeds have none. Probably this explains the anti-scorbutic value of sweet-wort and of beers made from lightly dried malt; and the total failure in this respect of our modern beers made from kiln-dried malt. Dr. Chick, amongst many other interesting and important results published by members of the Lister Institute Committee, states that the juice of raw swedes and of raw turnips is a valuable anti-scorbutic (to be added to milk for the use of artificially nourished infants); so, she states, is orange-juice. But, contrary to the usual opinion, she finds that beetroot has little or no anti-scorbutic value. The whole subject is of extreme importance, and is necessarily in a tentative stage of pioneer experiment.

www.ingramcontent.com/pod-product-compliance
Lightning Source LLC
Chambersburg PA
CBHW071414180526
45170CB00001B/102